Salmon farming handbook

Stephen Drummond Sedgwick

Fishing News Books Ltd
Farnham · Surrey · England

British Library CIP Data

Sedgwick, Stephen Drummond
  Salmon farming handbook.
  1. Salmon farming
  I. Title
  639.3'735

ISBN 0-85238-158-1

Published by
Fishing News Books Ltd
1 Long Garden Walk
Farnham, Surrey, England

Typeset by
Mathematical Composition Setters Ltd
Salisbury, Wiltshire

Printed in Great Britain by
Henry Ling Ltd
The Dorset Press, Dorchester

Contents

Illustrations and tables

5

Introduction

The members of the salmon family form the most valuable group of the world's fish species. They have been in the forefront of human interest wherever they occur. Isolated communities in the northern hemisphere have taken the greater part of their livelihood from these fish which migrate to the sea but must return to breed in fresh water. The west coast Indians in North America depended on runs of salmon homing to their local rivers for winter keep. Other tribes on the east coast fished for Atlantic salmon returning to rivers from Labrador to New Hampshire. Atlantic salmon in Scotland were put under protective legislation in laws given by William the Lion in the twelfth century. All kinds of political skulduggery have been used through the centuries and are still used to obtain fishing rights and commercially exploit wild stocks of both Atlantic and Pacific salmon. The almost mystical esteem given to these fish probably stems from their strange ability to find a way back to breed in their parent river after years of ocean wandering.

The Salmonidae are naturally distributed throughout most of the northern hemisphere, from the temperate zone northwards to beyond the Arctic Circle. There are no native salmon in the southern hemisphere but they have been successfully introduced into South America and Australasia. Nearly all members of the salmon family can adapt to life in salt water. Some species must migrate to the sea or die, most of the others have races which deliberately migrate. A few species have not adapted to life in the sea, mainly because they inhabit isolated freshwaters where they have become land locked and have no access to a saline environment.

The anadromous salmonids are all hatched and grow for a time in fresh water before they go down to the sea. They remain feeding and growing, some in the oceans far from shore, others in estuarial and coastal waters, until the onset of sexual maturity when an increasing urge to spawn sends them home to their parent rivers. The ability of salmon to find their way home back to the rivers where they were hatched and spent their parr life period is still a source of wonder to naturalists. The regularity with which different migratory members of the salmon family find their way varies to some extent between species but the majority return to the right river and some will go back to the parental gravel bed or even ascend the outfall from the troughs in a hatchery where they were artificially incubated.

Salmon have developed a sense by which to navigate the seas back to coastal waters. The long-distance migrants can return directly to a short length of coastline near their home rivers, from feeding grounds which may be thousands of miles away across the open oceans. It is difficult to believe that they can do this following only temperature gradients or currents. It is possible that they use the sun, moon and stars like some migratory birds, or have a means of sensing variations in the earth's magnetic field. The method by which returning migrants identify the home river or stream is now better understood. The young salmon become imprinted with the chemical characteristics of their freshwater environment. The adult fish, once close enough to recognize those characteristics, enter the river and smell their way upstream to waters near their original hatching and rearing ground.

Wild salmonids have been caught and stripped of their eggs, and the eggs artificially fertilized and incubated in hatcheries, since the middle of the nineteenth century. It was not until the early years of the twentieth century that serious thought was given to the commercial possibilities of not only breeding, but on-growing fish of these species as human food by domesticating and retaining them in captivity throughout their life cycle.

The Norwegians made the first move towards farming the sea because their freshwaters were too cold in winter and the growth season for the fish was too short. The sea temperatures along the north and west coast of Norway are on average higher than those in the rivers. It was known that the sea-going race of rainbow trout (steelhead) were naturally adapted to migrate to the sea when they had grown to a certain size in fresh water. In 1912 the Norwegian Storting approved the culture of rainbow trout in the sea but the original attempts at marine farming were a failure because the sea burst open the pens made to enclose the fish. The failure was the subject of some ridicule at the time and an article in a local paper suggested growing oysters which would not be so likely to escape if the barriers were washed away.

No further attempts of any consequence were made to grow salmonids in salt water in Norway until the mid-1950s. Sea farming then gradually increased over the next ten years and production reached about 500 tons in 1965. Methods initially used were generally a repetition of the original idea of enclosing a bay or an arm of a fjord with a fixed fence of netting. Some farmers tried pumping sea water to earth or concrete fish ponds on the shore but this proved to be expensive and was less profitable than other methods. Although the low cost of electricity in Norway and the very small

9

tidal differences kept pumping costs well below those in most other European countries, Norwegian sea farmers subsequently decided to concentrate on low-cost fixed tidal enclosures where they could find suitable sites. Now the use of floating net cages has superseded most other methods.

Farming fish of the salmon family in salt water in the British Isles started in 1960 at Loch Sween on the west coast of Scotland. The fish were first kept in net enclosures in brackish water and then transferred to floating cages in the sea loch. The trials were successful and further commercial development took place in the mid-1960s when a farm using fixed tidal enclosures was started in a brackish water loch on the Shetland Islands. At much the same time a shore-based rainbow trout farm using pumped sea water was started at Loch Ailort. This farm subsequently moved into floating cages in the sea and turned to producing Atlantic salmon.

The domestication of terrestrial animals, birds and mammals, followed by selection and selective breeding of the most successful species, races and types, has been practised throughout recorded history. Modern intensive fish farming most closely resembles the methods of production developed for broiler chickens, but development from the time the first red jungle fowl was lured out of the trees to the broiler bird has been a long process. We are at the very beginning of the domestication of salmon. No doubt, in the not too distant future, races or types of particular species of salmon, or hybrids between them, will be developed for the table market. Such fish may have a combination of desirable features that lend themselves to domestication and to intensive culture. Some progress has already been made in this direction. In the meantime it is proving profitable to use semi-wild fish. This involves looking at all the available species and races of salmon in a search for useful characteristics that lend themselves to farming in the sea.

The greater part of the surface of the world is covered by water. Mankind must eventually cease hunting wild fish and farm the waters as the land is farmed. Rational aquaculture may then involve the cultivation of plants and animals far lower down the food chain than the fish at present reared for human food. Fish of the salmon family are near the top of the food chain. When kept in captivity they have to be fed mainly on animal protein which is already in short supply. For the time being, however, they offer the best, and in many areas, the only economic return to fish farmers who farm fish in the colder seas of the higher latitudes.

1 The fish

Most salmonids, particularly those which are of
concern to fish farmers, share a similar basic anatomy
and physiology. The adults of the various species are
generally fairly easily recognized but inexperienced
observers may find the very young fish difficult to
distinguish.

Salmon have the typical arrangement of fins common
to most freshwater fish. The leading pair are the
pectoral fins on either side of the lower half of the
body. These are followed by a second pair of pelvic fins
on the ventral surface or belly of the fish and a single,
anal fin immediately behind the anus. There are two
fins on the back, the dorsal fin and a degenerated
adipose fin. The body ends in a caudal or tail fin. All
the fins except the adipose fin are supported by bony
rays.

The skin over the whole of the body surface, apart
from the head and fins, is protected by overlapping
scales and an outer coating of slime. The larval fish
when newly hatched have no scales and scale growth
does not start until the fish reach a length of about
3cm. Scale growth begins with the laying down of
platelets in the region below the adipose fin. The scales
lie partly embedded in pockets in the skin and grow

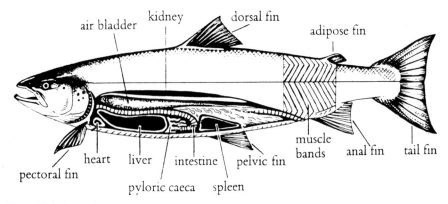

air bladder   kidney   dorsal fin   adipose fin

heart   liver   intestine   pelvic fin   muscle bands   anal fin   tail fin

pectoral fin

pyloric caeca   spleen

Fig 1   Main internal and external features of salmon

with the fish. They are cycloid and the outer, overlapping part is smooth.

Scale growth takes place at the periphery of each scale. Keratinous material is laid down in the form of visible, thickened rings on the embedded portion of the scale. The fish's body and the scales which cover it grow more quickly in summer than in winter. The parts of the scale surface where the growth rings are closer together or wider apart represent periods of winter and summer growth from which it is possible to read the age of wild fish. This is not so easy and is sometimes impossible in farm fish as growth can be more or less continuous and the different growth periods become indistinguishable.

The basic pigmentation of the body surface, green, brown, red or gold, is produced by colour cells or chromatophores in the skin. These cells have the ability of either releasing or withdrawing microscopic colour particles from the cell walls. This accounts for the fish's ability to change colour when against different backgrounds. Most pelagic fish including salmon in the sea are often partially or wholly silver-coloured along their sides. The highly reflective, silver coating is formed by microscopic, colourless crystals of a

14

metabolic by-product called guanin, which is laid down on the scale surfaces and in the surrounding chromatophores.

The body is supported by a flexible vertebral column which links together the head, body and tail fin. Propulsive power for swimming is provided by the main blocks of muscles on each side of the backbone which are supported by bony ribs. The thick, bony structures of the head protect the brain and specialized sense organs. Behind the semicircular canals, which are organs of balance on each side of the hind-brain or cerebellum, are cavities each containing an otolith. This is a small free-floating portion of hard calcium carbonate. The otoliths rest on hairy cilia lining the cavities and act as balance mechanisms by indicating body aspect to the fish's nervous system. Otoliths grow with the fish and when ground thin and polished show dark and light rings corresponding to winter and summer growth periods.

The heart is at the leading end of the body cavity, behind the head and below the vertebral column. The body cavity extends back to the anal opening which is between the leading ends of the pelvic fins. Food passes through the oesophagus into the stomach to the distensible walls of which are attached a number of blind appendages called pyloric caeca. The digestive organs include a liver, gall bladder and spleen. Digestion continues in the intestine and undigested material is discharged from the anal opening as faeces. The fish have a kidney situated along the underside of the backbone. The kidney not only eliminates waste products but also serves an important osmoregulatory function, controlling the salt to water balance in the fish's body fluids.

The swimbladder is a specialized organ of balance situated in the body cavity below the kidney and above the gut. In some species it can be filled either by gulping air at the surface or through absorption from the bloodstream. The extraction of air from the blood

15

can take place as an automatic reaction to adjust the specific weight of the fish in relation to the pressure at the swimming depth.

The gonads or sex organs are also contained in the body cavity. The cell structures giving rise to eggs or sperm are not initially differentiated in the larval fish. On the approach of sexual maturity the developing gonads occupy an increasing proportion of the body cavity and in the later stages of development fill the space to an extent which impedes the intake of food to the stomach and partially inhibits the digestive processes.

Respiration An exchange of dissolved oxygen and carbon dioxide between the water and the fish's blood takes place through the gills. The gill filaments are carried on branchial arches in cavities on either side of the head which are open to the oesophagus on the inside and covered by a gill cover or operculum on the outside. A uni-directional flow of water over the gills is maintained by the branchial pump. With the gills closed, the mouth of the fish opens and water is sucked in, filling the mouth and the buccal cavity which is formed by the forward end of the oesophagus. The mouth is then closed and water is passed through the gills and out of the opening gill covers. The movements of the parts of the branchial pump are continuous but the rate of opening and closing varies with the fish's demand for oxygen. This in turn can depend on whether it is stressed in any way or is stimulated to greater activity. An increase in breathing movements in farm fish is often an indication of stress and is most usually due to lack of oxygen in the water.

The gill filaments are densely supplied with blood vessels. Carbon dioxide is released to the water at the same time as the oxygen is extracted which is then transported in the red cells through the fish's body. The main source of oxygen in water is the atmosphere

16

but the photosynthesis of green water plants can be of great importance to fish and to the fish farmer. The production of oxygen by plant metabolism makes a considerable contribution to the total dissolved oxygen in the water. This ceases at night and can cause a significant fall in the oxygen content of the water.

The oxygen content of water is measured in milligrammes per litre. A level of 6mg/l can, for practical purposes, be taken as the lower limit for salmon. A level of 8–10mg/l is satisfactory for on-growing and 10–12mg/l for hatcheries and fry or smolt production. Cold water can take up and hold more oxygen and the amount which remains in solution becomes progressively less as the temperature rises. The percentage of saturation is also a way of representing oxygen content. Fast-flowing, cold water can be supersaturated with oxygen to more than 100% saturation. The lethal level of oxygen depends to some extent on the level in the environment to which the fish have become acclimatized. This is most likely due to an increased ability to extract oxygen from water with a low concentration. The oxygen content of water can be measured quite simply in practice using an electric meter. This is a small, portable piece of apparatus which can either be read directly or connected to a continuous recorder.

Oxygen demand The fish's need for oxygen varies with activity, temperature and food intake. Fat fish in good condition need more oxygen than thin fish. Young fish (fry, parr and smolt) need proportionately more oxygen while growing than when they are approaching maturity.

An ample supply of oxygen is particularly important during the incubation of eggs. The metabolism of the developing embryo is very demanding of oxygen and growth may be slowed down or the embryo suffocate if oxygen intake is inadequate. The enhanced need for

17

oxygen continues during the alevin stage while the larval fish remains dependent on the yolk sac. The most dangerous time, when oxygen demand is greatest, is just before hatching. One of the dangers of artificially increasing the water temperature in a hatchery, in order to accelerate development and reduce incubation time, is that the developing embryo may be unable to meet the increased demand for oxygen at the higher temperature, because insufficient oxygen can pass through the egg membrane.

The increase in oxygen consumption due to activity resulting from swimming in order to maintain position in a current of water or to capture food is directly proportional to the swimming speed, and is approximately similar for all salmonids of interest to fish farmers. For example, the oxygen consumption of a fish cruising at a speed of 30cm per second is about half that for one swimming at a speed of 75cm per second, or a fish having to maintain station against a flow of water of the same speed. This is important for fish farmers to understand as it affects not only the amount of oxygen the fish will need in tanks or raceways with fast-flowing water, but also their consumption of energy in the form of food.

Increasing temperature not only reduces the amount of oxygen which is carried in the water, but also increases the demand for oxygen in the fish. This is primarily due to spontaneous metabolic activity in response to increasing temperature. The oxygen requirement is approximately five times greater at 20°C than it is at 5°C. This is of vital concern to the fish farmer as it means that the density at which their fish can safely be kept becomes progressively lower as the water temperature increases. The increase in the fish's demand for oxygen due to spontaneous activity is almost linear between 5°C and 20°C in farmed salmonids and it is possible to work out the proportional density at which the fish can safely be kept at a given water temperature. If the fish start to

18

feed, this induces an additional, temporary demand for oxygen. This is not of any great significance at temperatures below 15°C but at higher temperatures the density at which the fish are kept determines the maximum temperature at which they can safely be fed.

In sea water of oceanic salinity and a temperature range of 5°C–15°C, the density at which fish can be safely kept decreases by approximately 25% for each 2·8°C rise in temperature. For example, a cage containing on-growing Atlantic salmon in the sea at a density of 10kg per cubic metre at 15°C could safely hold about double that weight of fish at 5°C. In practice, the density at which on-growing fish should be kept is the weight per cubic metre corresponding to the optimum temperature for growth. Less food or no food should be given at times when the water exceeds this temperature. The density can be proportionately increased when fish are over-wintered at lower temperatures.

Circulation   In the circulation of fish, blood is pumped by the heart, which has four chambers connected in series separated by valves, to the gills where the exchange of respiratory gases takes place. It then flows back to the heart through the systemic arteries and veins serving the parts of the body and the separate organs.

The heart rate accelerates when the fish are stressed. It also increases with rising temperature as part of a general spontaneous increase in metabolic activity. This can be particularly damaging to the developing embryo or to alevins in the yolk-sac stage when the pulse rate can double if the temperature rises from 5°C to 12°C.

The volume of blood is relatively low (about 2·5% of body weight) and much lower than in mammals. It contains white cells (lymphocytes and leucocytes) and red cells (erythrocytes). The red cells are fewer in number and larger than those in mammalian blood, and contain less haemoglobin.

19

Digestion and excretion   The digestive tract of salmonids can be separated into two main regions. The forward part which is made up of the mouth and buccal cavity, and the oesophagus, stomach including the pyloric caeca, and the intestine, together with their associated ducts to the liver, pancreas and gall bladder. The prey is caught and held by oral teeth and is often swallowed whole. The oesophagus is a short, thick-walled tube which may serve to reject as well as accept food before it is passed into the stomach. The function of the stomach is to break down food into soluble particles which can then be absorbed through the gut walls. The secretion from the gastric mucosa is acid and contains pepsin as well as other enzymes. The nutrients enter the bloodstream as soluble proteins and fats together with a proportion of carbohydrates.

The time taken for the food to pass through the body and for undigested matter to reach the anal opening and be excreted varies with the water temperature. The metabolic rate increases as the temperature rises to the optimum for food conversion. The excretion of waste materials from the gut takes place through the anus in the form of faeces.

Gill cells also play a part in the uptake and removal of salts and the elimination of waste products. They function with the kidney in maintaining osmotic equilibrium between the body fluids and the environment. The kidney tubules act as filters. Urine is formed by glomerular filtration and the renal excretion is passed from the body by ducts near the anus.

Osmotic regulation   The blood of fish in either fresh or salt water must undergo continuous changes in order to maintain a balance between the salts in solution in the body fluids and the surrounding water. The process by which this is carried out is known as osmosis. When solutions of different concentration are separated by a semipermeable membrane, water will pass through the

20

membrane from the dilute to the more concentrated solution until the concentration on both sides of the membrane is the same. The body fluids of fish in fresh water are more saline than the environment. The gills, the gut and to a lesser extent the skin, are semipermeable membranes. Water enters the bloodstream through these membranes and must be constantly discharged through the kidney to maintain the correct saline balance in the body fluids.

In a marine environment, the sea water is a more concentrated salt solution than the body fluids and water passes out through the semipermeable membranes and is lost to the body. The fish must drink sea water to compensate for this loss and in doing so take in water that already contains more salt than their body fluids. The marine teleost fish are adapted to cope with having to swallow sea water to prevent dehydration caused by the osmotic loss of body fluids.

Salmon and other anadromous salmonids undergo a special metamorphosis known as smoltification which enables them to adapt without undue stress to the change from fresh to salt water. The few species and races of salmonids which do not naturally migrate to the sea at some stage during their life cycle are not truly stenohaline, which means that they can only tolerate small changes in the salinity of their environment, but can adapt by acclimatization to life in salt water.

Fish in a freshwater environment have no special need to retain water in order to maintain a stable hypertonic solution in their body fluids. Water containing a relatively small amount of dissolved salts is taken into the body fluids through the skin. Water and salts are excreted in the urine and a very small amount of salts through the skin.

Fish in the sea must retain water and excrete salts to maintain the hypertonic balance. They take in salt water and excrete salts (and the minimum of water) through the gill cells. Salts also enter the body fluids from the gut and water passes out through the skin. The saline balance of the body fluids is principally maintained by the salt-excreting activity of gill cells assisted by excretion of salts in the faeces. The urine contains less water and more concentrated salts and less water is lost through the skin. Carnivorous fish also obtain significant quantities of water of low salinity in the hypertonic body fluids of their prey.

The most obvious physiological change which smoltifying salmonids undergo is the laying down of a silvery coating of guanin in the skin. A function of this crystalline deposit appears to be to act as a barrier to osmotic exchange and to prevent the loss of water through the skin. At the same time, prior to the fish entering the sea, the specialized basal cells in the gills proliferate and increase the fish's ability to excrete salt.

Reproduction Sexual maturity in fish involves the mobilization of large quantities of material for building specialized tissue and food to store in the developing eggs. In the male fish the collection of tissue-building material is not only needed for the developing gonads but also to form the structural sex characteristics such as the kype or protrusion on the lower jaw and the hump on the back of pink salmon. Anadromous salmonids all undergo a period of starvation before they reach their spawning grounds and must store sufficient fats, protein and carbohydrates not only to create eggs and milt but also to survive until they are ripe to spawn. The greatest demand is for fats, and the fat content of some species of salmon is reduced by 85% between the start of spawning migration in the sea and arrival on spawning grounds in the head-water of a river. The body fats carry the carotenoids which colour the tissues pink or red and as they are used up the flesh becomes pale, watery and tasteless. It is this loss of stored nutrients which renders sexually maturing or mature salmonids worthless to the sea farmer who must either stave off maturity in his fish or slaughter them before it destroys their value.

The process by which eggs develop in the ovary of the female fish is known as oogenesis. Maturation of the gonads is stimulated by hormonal activity and may take place over a variable period governed by racial factors as well as being characteristic of a particular species. The eggs develop in a pair of ovaries and when ripe are released and collected in the abdominal cavity. The eggs are covered by a soft shell in which there is a small opening or micropyle below which is a spot of protoplasm surrounding the nucleus. The male testes are also paired and in them the process of spermatogenesis produces the mobile, free-swimming sperm cells. The sperm must find and enter the micropyle in an egg in order for fertilization to take place.

Salmonid sperm and eggs are shed together by the male and female fish. Fertilization takes place at once because the micropyle starts to close as soon as the eggs enter the water and begin to swell and harden. Sperm can only penetrate the eggs for a very short time before it is too late and the micropyle is closed. Initially the eggs are slightly adhesive and this continues during the period water is taken up and the eggs swell and harden completely. The cell cleavage begins during this period.

The developing embryo subsists on protein, fats and carbohydrates stored in the yolk. The egg shell or chorion has the ability to pass in oxygen and some essential nutrient salts from the water and to excrete waste materials. The time taken for embryonic growth varies with species and with water temperature. The demand for oxygen is high throughout incubation and highest just before hatching. The characteristic orange, oily fat globule remains in the yolk sac after the larval fish have hatched and continues more or less unaltered

23

through the alevin stage being finally used up just
before the fry begin to feed.

Life in the water Good husbandry in fish farming depends to a great
extent on appreciating that fish are aquatic not
terrestrial animals. They are poikilothermic (having a
variable body temperature) and depend for their well-
being on a relatively stable thermal environment,
which is quite unlike that shared by human beings
with their other domestic animals. It is necessary to
know and understand something of the special
behavioural adaptations that equip fish for their life in
the water.

Sight and visual The large eyes of fish differ from those of mammals
stimulation and in some ways are more efficient. The mammalian
eye focuses by muscles altering the shape of the lens.
The whole lens in the fish eye moves backwards and
forwards like the lens in a camera. The cornea has the
same refractive index as the water and the pupil bulges
outwards taking in a wide visual field in both the
horizontal and vertical planes. Although the eyes are
set on the sides of the head, the field of vision overlaps
in a forward direction giving stereoscopic sight over a
cone of about 25 degrees. Theoretically, in completely
still water, upward vision through the surface is
restricted to a cone of a little less than 100 degrees.
The size of this circular porthole varies with the depth,
and beyond the periphery the fish sees only a reflection
of the bottom. In reality, water is seldom motionless
and fish swimming near the surface get glimpses in
many directions through the waves.

Salmonids are thought to see fairly well in air and
also appear to have good vision in semi-darkness. The
response to visual stimuli is immediate. The reaction to
sudden changes in light intensity and to observed
movement is much more violent when the fish are in a
confined area from which they are unable to escape.

24

The stress suffered can seriously interfere with the normal metabolic processes. Most of the salmonids used for fish farming are still half wild and it may be necessary to protect them from stress resulting from visual stimulus. Domesticated stocks need no protection as they learn to respond positively to land movements which they associate with the delivery of food. Young Atlantic salmon are highly sensitive to visual stimuli and are easily stressed during their freshwater life. Rearing tanks should be kept shaded and food given by automatic feeders.

Lateral line The division which can be seen along each side of the body of salmonids and most other fish is a sensory organ, passing through holes in a row of specialized scales. The organ has branches opening to the surface between each scale and a group of nerve-endings connected to a nerve running along below the scale pockets. It is thought that changing pressures resulting from the fish's own movements, either when swimming or at rest, register a basic pattern in the central nervous system. Any additional changes in pressure, even though very small, upset this pattern and inform the fish of the shape and direction of the disturbance. The lateral line openings are thought to be a means by which the fish can maintain their position, avoid danger and search for prey in conditions when they are deprived of vision. Farmed salmonids may react positively to splashes on the surface and learn to associate this with the arrival of food, but it is probably more important to avoid, as far as possible, any disturbance in the water as this could be a source of stress.

Hearing and balance The salmonid ear is similar to that of other vertebrates and is made up of an inner ear and labyrinth which function as organs of balance as well as hearing. The labyrinth consists of three fluid-filled, semicircular

25

canals, each with a separate ampulla. Below the
semicircular canals are capsules containing otoliths
made of hard calcium carbonate whose main function
is the perception of gravity and the maintenance of the
fish's equilibrium.

Research carried out on minnows and some other
species of freshwater fish indicates that they respond to
sounds with a frequency of about 6,000–7,000 Hertz (1
Hertz = one wave per second). Sound travels more
quickly in water than in air and has been used as a
means of attracting fish, free-swimming in a large area
of water, to a central point for artificial feeding. The
sound used is within the ordinary, musical range and is
neither extra high or low frequency. Farmed fish do
not appear to be stressed by sound within the human
auditory range, but waves of low frequency in a range
of 5–25 Hertz may be sensed by the lateral line,
causing stress and an endeavour to escape.

Swimming and resting Fish in a current of water are obliged to swim in order
to maintain station. In static water, they only require
to swim voluntarily in the pursuit of prey, to avoid
predators and to keep contact with other fish of their
own species. Salmonids use only the caudal or tail fin
for swimming which is flexed in a sculling motion by
the blocks of muscles along the sides and so propels
the fish through the water. The dorsal and anal fins are
mainly used for vertical balance, and the pectoral and
pelvic fins for lateral balance and when resting.

The swimming speed of salmonids is proportional to
their length, but it is very difficult to determine
accurately either the idling or maximum speed of any
size or species of fish. Several different salmonids, both
adult and juvenile, have been tested in flow channels
where the opposing current can be progressively
increased but it is impossible to tell whether it is
fatigue which causes the fish to fall back or turn and
swim downstream or other reasons. Certainly

26

anadromous salmonids are capable of rapid acceleration over short distances but it is doubtful how long or for how far these bursts can be sustained.

Farmed fish kept in raceways or tanks with a concentric flow must swim to maintain station. In so doing they are using up energy sources which would otherwise have been stored in growing body tissues. Farmed salmon do better in terms of food conversion and growth rate in static water where there is no directional flow, but they need a greater volume as living space in order to obtain sufficient oxygen and have to be kept at a reduced density.

2 The Pacific salmons

There are six species of salmon which are native to the
rivers flowing into the Pacific and eastern Arctic
Oceans. Five of these species are native to rivers in
North America. They also occur naturally in rivers on
the Kamchatka peninsula in the USSR which flows
into the sea of Okhotsk and in other rivers running out
into the Bering Sea and to the far north into the east
Siberian and Laptev Seas. The sixth species is a small
salmon called masu which runs up the rivers on the
islands of Hokkaido in Japan and Sakhalin off the coast
of Manchuria. It may also occur in some rivers on the
Manchurian mainland.

Two species of Pacific salmon have been successfully
introduced into rivers in the southern hemisphere and
more recently into rivers in the European Arctic
flowing into the White Sea from the Kola Peninsula to
the west of Murmansk.

In common with most other members of the salmon
family some Pacific species have races which naturally
spend their entire life cycle in fresh water and others
have been successfully acclimatized to freshwater life,
notably in the Great Lakes in Canada. They can grow
to full size without going to sea provided there is
sufficient food in the freshwater environment.

Successful growth on free range in freshwater lakes depends on the presence of a shoaling food fish in sufficiently large numbers, usually belonging to the smelt or herring family. The freshwater populations which have been deliberately introduced have so far had to be supported by hatcheries.

The freshwater introductions in Canada were originally made in order to find a substitute species for the Great Lake trout (*Salvelinus namaycush*) which is in fact a char. This species, which was of great commercial importance, had been decimated by the colonization of the Great Lakes by the sea lamprey (*Petromyzon marinus*) which followed the opening of the St Lawrence seaway. The lampreys spawned in the rivers draining into the lakes. Their progeny no longer needed to migrate to the sea as they could grow to maturity feeding on the larger freshwater fish species. The lampreys were eventually controlled by using a specific poison. The lakes were then stocked with two species of Pacific salmon. These introductions have proved enormously successful, not only in providing the opportunity for a commercial fishery of much greater value than that for the lake trout, but in the contribution the new fishery has made to sport angling in both Canada and the USA.

The five most important species of Pacific salmon belong to a single, separate genus and they are distinct from Atlantic salmon, the trout and the chars. The generic name of the Pacific salmon is *Oncorhynchus* and this precedes the proper name of each species. The individual species have many different common names. Those used here are generally adopted by the salmon fishing industry in Canada and the USA. They are pink (*gorbuscha*), sockeye (*nerka*), coho or silver (*kisutch*), chum (*keta*) and spring or chinook (*tschawytscha*). The Latin names sound strangely exotic because the species were originally classified by Russian scientists working on rivers on the opposite side of the ocean.

29

ATLANTIC SALMON

1 Jaws and eye line
2 X-shaped spots
3 Small tail stalk or wrist
4 Distinctive shape of tail fin

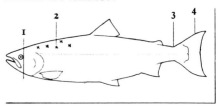

PACIFIC SALMONS

1 Fine speckles
2 Narrow tail stalk or wrist
3 Distinctive shape of tail fin

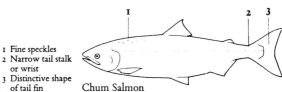

Chum Salmon

1 Fine speckles
2 Distinctive shape of tail fin

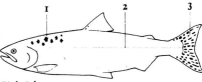

Sockeye Salmon

Fig 2 Distinguishing features of salmon

1 Very large oval thumbprint-sized spots
2 Large number of scales in median line series
3 Large spots on tail fin

Pink Salmon

1 White gums
2 Medium-sized spots
3 Tail only slightly forked with spots on upper lobe only

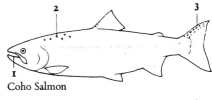

Coho Salmon

1 Black gums
2 Large spots
3 Spots on dorsal and tail fins

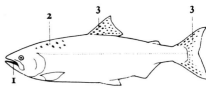

Chinook Salmon

30

All five species exist in the wild in great abundance. They have been fished for as a staple food by man since prehistoric times. The natives on both sides of the North Pacific originally fished for salmon on their return to spawn in their parent rivers. More recently they have been fished for further afield, on their ocean feeding grounds, using sophisticated fishing gear and drift nets. The inroads made on stocks by increasingly efficient fishing methods have led to the imposition of greater controls, firstly on a national and then on an international basis. Now the catch and spawning escapement in North America is scientifically regulated and a measure of stock management has been achieved.

Recent developments have seen the increasing success of ranching techniques being employed in Japan and in the USA. These involve rearing young salmon in an artificial environment, releasing them to feed, free ranging in the sea, and recapturing them commercially on their return to the place where they were released.

All five species must return to fresh water to spawn. The young of two of the five species go down to the sea as fry, either shortly after the yolk sac has been absorbed or early in the first summer. The other three species have a parr life period of one or two years in fresh water and migrate to the sea as smolt.

The marine diet varies to some extent between the species but consists principally of euphausid shrimps, and the fatty shoal fishes such as pilchards, anchovies, eulachons, smelts and the needlefish (*Cololabis saira*). The species which migrate as smolt feed initially on a particularly abundant euphausiid shrimp (*Thysanoessa spinifera*). This food animal, which is known as the 'red feed', is the main constituent of the diet of sockeye salmon throughout their sea life. This species is less piscivorous than the other four and does not feed, to the same extent, on other fish.

Some species meet on their oceanic feeding grounds and the eastern race of pink salmon from the

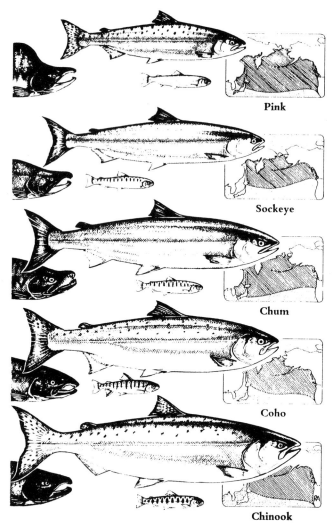

Fig 3 Pacific salmon:
typical ocean-going adults,
fry and heads of spawning
males. Maps show sea
feeding grounds; stocks
from both sides of the
Pacific Ocean may meet in
the central area

Pink

Sockeye

Chum

Coho

Chinook

Kamchatka rivers mingles with the pink salmon from
North American rivers feeding in the sea. The marine
life before reaching sexual maturity also varies between
the species. Pink salmon are the smallest and the
youngest at the time of return on spawning migration,

32

having spent one or one and a half years at sea.
Sockeye salmon spend two or three years growing to
sexual maturity and chum and chinook disappear into
the ocean for three or four years. Coho salmon more
closely resemble Atlantic salmon and spend one or two
years at sea returning to spawn, when they are between
three and five years old.

Chum salmon
(***Oncorhynchus keta***)

Chum salmon occur in Asia and North America and in
Japan. They are most abundant in the Asian rivers.
The common names are *keta* in Russian and *sake* in
Japanese.

Recognition

The body is elongated and rather compressed when
viewed from above. The caudal peduncle (junction
between body and tail fin) is slender. The head comes
to a point at the mouth which has well-developed teeth
that turn into fangs when the male fish are in spawning
dress. The first gill arch has 10–16 rakers and there are
150–160 scales counted along the lateral line. The
scales are similar to those of other salmon and show
the age of the fish.

The colour in marine feeding dress is metallic blue
on the back with a few black flecks. There are no
regular black spots. The fins are edged with black
particularly in the male fish. The appearance
completely changes when the fish shift into spawning
dress in fresh water. The males develop distinctive
dusky red streaks on their sides which join together
below the lateral line and the body colour turns a
greenish-yellow along the sides of the fish. The heads
of the male fish have the elongated and slightly dished
appearance typical of the salmon family but the leading
end of the lower jaw does not develop the pronounced
upward growth or kype typical of Atlantic salmon.

The flesh colour of the fish in marine feeding
condition is pale pink and the fat content is 9–11%.

The young fish have green iridescent backs and parr

33

finger-marking on their sides which disappear just below the lateral line.

Spawning The majority of fish reach their spawning grounds between September and January but in some northern rivers spawning can take place in June or July. The spawning process is similar to that of Atlantic salmon and takes place in fast-flowing, shallow streams with clean gravel beds. The process lasts over three to five days. In very cold regions the female fish appear to select gravel where there are springs in the river bed which maintain a temperature that does not fall below about 4°C.

The number of eggs shed varies between 2,000 and 5,000. The eggs are comparatively large and average about 7mm in diameter. Both males and females of all the species of Pacific salmon die after spawning.

River life Eggs hatch after 100–120 days. The alevins average 23mm in length and this phase lasts 30–50 days dependent on water temperature. The larval fish start to feed before the yolk sac is fully absorbed.

The fry emerge from the gravel some time between March and May depending on the location. They can remain in the river for several months but the majority leave for the sea early in the summer after a few weeks of freshwater life.

Chum salmon fry show an increasing preference for salt water and become pre-adapted to marine osmoregulation. The fry are very difficult to maintain in fresh water for any period after their normal time of migration.

The average size at migration to the sea is 3·5cm.

The food in fresh water consists of plankton animals followed by insect larvae.

Marine feeding and growth in the sea Marine food initially consists of zooplankton followed by copepods, euphausids and other small marine animals including some small fish. The diet contains a

34

high proportion of invertebrates throughout the period of sea feeding.

Most fish reach sexual maturity after three years marine life but some spend only one year in the sea and others as many as six years before returning to their parent rivers on spawning migration.

The average length of the fish when three years old is 50–70cm and the average weight 2·5kg. The largest fish, which are six years old, can be up to 6kg in weight.

Hybrids   Chum salmon have been successfully crossed with pink and sockeye salmon and the hybrids back-crossed with other hybrids within these species or with the species themselves.

Fish farming   Chum salmon have not attracted much attention as a species for farming. The reason is most probably that they were not considered a very high-quality fish in North America when compared to the other Pacific species and would cost as much to feed-on to market size in captivity.

The procedures for incubation and hatching are straightforward and chum salmon are now being 'ranched' on a very large scale in Japan. The fry are hatched artificially and released to feed on 'free-range' in the sea. The crop is partly taken by enhanced off-shore marine fishing but the greater proportion of the catch is made by traps as the fish return on spawning migration.

Chum salmon are also ranched in the USSR and have been introduced into the rivers draining into the White Sea. Chum salmon spawned in these rivers have been caught in Norwegian waters.

Sockeye salmon   It can be argued that the differences in the behaviour
(***Oncorhynchus nerka***)   between the various species of Pacific salmon have developed as a means of avoiding competition and in

35

order to make maximum use of the available freshwater environment for reproduction. Some species have individual races with widely differing patterns of behaviour involving time of return from the sea on spawning migration, habitat during parr life and adaptation to a completely freshwater life.

Sockeye salmon consist of many different races not only present in different rivers but also occurring in different parts of the same river system. The species also has a race completely adapted to spend its entire life cycle in fresh water which occurs in many lakes in British Columbia.

Recognition The body is less elongated than in chum salmon, which are not unlike sockeye when feeding in the sea, but the species can be easily distinguished from all the other Pacific species by the number of gill rakers (28–40) on the first gill arch.

The fish are bright silver when feeding in the sea and the body and fins are unmarked apart from some fine speckles on the back which is shaded blue-green. In spawning dress the bodies of both males and females turn deep red and the heads become dark greenish-grey. The males develop pronounced protrusions on the ends of both the upper and lower jaws.

The flesh during marine feeding is deep red and when canned is regarded as the highest grade.

Sockeye during early life before migrating to the sea have typical parr markings which become obscured during smoltification. Races which remain and grow to sexual maturity in fresh water become silvery but are more spotted than the sea-going form.

Spawning Sockeye salmon usually spawn in small tributary streams above a lake, often having made a long and arduous journey upstream. Vast numbers, perhaps 15 million spawning pairs, are concentrated in particular parts of the range of the species, notably the

36

Frazer River in British Columbia and rivers running
into Bristol Bay in Alaska. Generally the early
returning fish travel furthest upstream. The first runs
into the Frazer River start in July and early August
and continue into late September and October.

River life   The young fish hatched in tributary streams drop back
to the lakes to feed on zooplankton followed by the
typical freshwater diet of young salmonids consisting
mainly of insect larvae and small crustaceans. A small
proportion of sockeye migrate to sea before they have
completed one year of freshwater life but the majority
spend one or two years as parr. The proportion of
larval fish spending more than two years in fresh water
before migrating to the sea increases in the stocks
belonging to the Alaskan rivers.

Freshwater races and   The process of smoltification and adaptation to marine
adaptation to salt water   life is similar to that of other anadromous salmonids.
Races of sockeye can remain throughout their life cycle
in their freshwater lake habitat although they are freely
able to go to sea. The freshwater races are called
kokanees. They feed round the lakes in concentrated
shoals and do not grow to a weight of more than about
500–750g. Kokanees have not so far been farmed for
the table market but they have been extensively
stocked as sport fish for angling in many lakes in the
western USA.

Marine feeding and   Sockeye are the least piscivorous of the Pacific salmon
growth in the sea   and their marine diet consists mainly of small crustacea
but includes some small fish.
The fish reach sexual maturity after two or three
years feeding in the sea. The age at spawning
migration can be a racial characteristic but the typical
four-year cycle of Frazer River sockeye, consisting of
one year in a freshwater lake, three years in the sea
and a return to the parental spawning ground to spawn
and die, can vary considerably between fish of the

37

same stock. Some fish may return a year older having spent an extra year in the sea and others a year younger having migrated from fresh water in the first summer of larval life. The proportion of five-year-old and even six- or seven-year-old fish increases in the Alaskan rivers.

The largest fish can grow to a weight of 6–7kg and a length of 75–80cm. The average canning weight is generally about 2·25kg and the fish are generally small in relation to some of the other Pacific salmon.

Fish farming   Although sockeye salmon eggs can be incubated and hatched without difficulty and there would seem to be no special problems connected with on-growing, they have not so far been of much interest to fish farmers. Their excellent flesh quality and the fact that certain races mature late seem useful attributes and the natural ability to adapt to life in either fresh or salt water also seems an attractive characteristic.

Pink salmon (***Oncorhynchus gorbuscha***)   This species is sometimes called humpback salmon because of the ugly cartilaginous 'hump' that forms on the backs of male fish when they are approaching spawning time. It occurs in large rivers in North America and Asia and is the most abundant species in the Alaskan salmon fishery and in the North Pacific. The Russians have introduced pink salmon into rivers running into the White Sea and the species now spawns sporadically in the lower reaches of rivers in north Norway.

Recognition   The first gill arch has 28 rakers. The fish can be distinguished from all the other Pacific salmon species by the large number of small scales. It has 170–240 in the first row above the lateral line.

The overall colour in the sea-life period is silver in both sexes with greenish-blue along the back. The fish are generally distinctively marked with a few large,

38

dark oval marks (about thumb print size) on their backs and on the lobes of the tail. The male fish become brick red in spawning dress as well as developing the characteristic 'hump'.

The flesh is pink, rather than red, which accounts for the common name of the species. They are regarded as being of lesser quality for canning than sockeye salmon.

Spawning

There are many different races of pink salmon with minor differences in behaviour patterns, but in most rivers where the species is native spawning takes place in late summer (August–September) not far above the head of tide. The eggs are shed in nests or redds made by the female fish in fairly shallow, fast-flowing water, typically favoured by salmonids. The average female fish produces about 1,500 eggs.

River life

The eggs take approximately 100–125 days to hatch and the alevins (newly-hatched young fish) have exceptionally large yolk sacs. The alevins remain passive in the gravel until the water begins to warm up in the spring which can be as early as March in southern rivers and well into May in northern rivers. The long time which the alevins remain in the gravel makes them vulnerable to suffocation by silting. The fry are silvery when they leave the gravel, without the spots and parr markings of the other salmon fry, and they descend directly to the sea.

Marine feeding and growth in the sea

During the first months of sea life the young fish collect in dense shoals close inshore before moving out to their ocean feeding grounds. The fish feed on crustaceans, squid, sandeels and other small fish. Marine growth is rapid and pink salmon reach sexual maturity in two years. The short life cycle distinguishes this species from the other Pacific salmon and also from other anadromous salmonids. The weight at maximum sea-growth is 2–3kg and the

39

length 40–50cm. A few fish can grow to a weight of 4·5kg in their short, two-year life span.

There are numerous races of pink salmon and the year classes of different races go through regular cycles of abundance. The fish produced in years of low survival are generally larger than those from the year classes with a high rate of survival to maturity.

Fish farming Pink salmon have been reared to maturity in captivity in Europe as well as in North America and Asia. They have proved a useful species to cultivate in sea cages or enclosures.

The prolonged period of alevin development can be shortened if warm water is available but great care must be taken not to over-accelerate the naturally slow absorption of the yolk sacs. A supply of salt water, preferably of full oceanic salinity is essential, as the fish are too small to be transferred to netting cages, even of the smallest mesh, until they reach a weight of about 200 to a kilogram.

Pink salmon can be grown to market size in captivity in less time than they take in the wild, using ordinary high-fat salmon diets. The wild races can be farmed successfully, but in the longer term selective breeding or hybridization may lead to the development of a more truly domesticated sub-species.

Coho salmon
(*Oncorhynchus*
***kisutch*)** Coho or silver salmon are the Pacific species whose behaviour and life cycle most closely resembles that of Atlantic salmon. They are also a sport fish for anglers. They are strongly piscivorous and are fished for commercially by trolling with spoons or spinning baits. The species readily adapts to life in fresh water and grows to full size provided there is an adequate supply of food fish.

Recognition The internal feature which distinguishes the species from the other Pacific salmons is the low number of

40

pyloric caeca (45–83). Externally, the tail is less forked than in the other species.

The marine feeding colour is silver with small black spots on the back and on the lobe of the tail. The gums are noticeably white at the base of the teeth. The male fish become red in spawning dress and develop a kype or hook on the lower jaw.

The flesh is pinkish-red in colour, very similar to that of Atlantic salmon. Large quantities of coho eggs are preserved in brine and sold as 'red' or salmon caviar. Unlike the commercial catch of the other Pacific salmon which is generally canned, coho are usually marketed fresh and quick-frozen.

Spawning   Coho are the salmon of the lesser rivers. They make their way into the smaller streams to reach their spawning grounds, which may be close to the head of tide or far up in the headwaters. Spawning behaviour is similar to that of Atlantic salmon.

Female fish can vary considerably in size. The quantity of eggs produced is between 1,500 and 2,000 per kg of body weight. Spawning takes place in the autumn or early winter, earlier in the more northern rivers. Coho do not deteriorate in condition on spawning migration to the same extent as the other Pacific salmon and some spawned-out fish remain alive for several months, although they all die before the spring.

River life   The hatching of eggs is controlled by the water temperature and incubation takes approximately 400 day/degrees. The fry usually emerge from the gravel in late April or May, earlier or later depending on location north or south in the range of the species.

The parr have brown backs and orange fins. Their behaviour, freshwater growth and size at seaward migration resemble more closely Atlantic salmon than the other Pacific species. Coho parr remain in fresh water for one or two years before smolt migration. The

41

average weight at smoltification in one year is about 20g.

Marine feeding and growth in the sea

Coho migrate long distances in the sea and their marine feeding behaviour is very similar to that of Atlantic salmon. They spend the first year mainly feeding on crustaceans before becoming almost completely piscivorous. They then feed on high-fat shoal fish such as herrings and needlefish. Growth is very rapid in the second year and the weight increases from 1–1·5kg in March to 5–7kg in September–October when they start on spawning migration.

The life cycle of most coho consists of one year's freshwater parr life followed by two years in the sea. Parr life can extend to two years and adult fish may not return to spawn until they have reached a total age of three to five years.

The average weight of fish at the end of the marine feeding period is normally between 2·5kg and 6kg and the length between 60 and 90cm according to the time spent in the sea.

Fish farming

Coho is the Pacific salmon species which has attracted most attention from fish farmers in North America and is of increasing interest for sea farming in Europe.

Coho salmon reared in floating net cages in the Puget Sound area of Washington State grew to a weight of 340g in six to eight months. A considerable proportion of the coho farmed in the northwest USA are slaughtered and marketed at what is termed 'pan-size' when they have reached a weight of 250–300g.

The fish that are grown-on may reach sexual maturity after ten to twelve months in the sea when they weigh between 1 and 2kg, as opposed to wild fish in the same area which weigh between 2·5 and 5·5kg when they return on spawning migration. The tendency to mature early may be an additional inducement to slaughter the fish as soon as they reach marketable size.

Initial growth in fresh water can be accelerated if the temperature is raised to between 11°C and 12°C and pre-smolt coho parr can be grown to a weight of 18–20g ready for transfer to salt water in seven to eight months compared to 12–14 months in unheated water. The total growing time taken from fry to slaughter at 'pan-size' can therefore be reduced to about fourteen months.

An interesting feature of the natural behaviour of coho is that they can grow to maturity in fresh water without being transferred to the sea. This factor will no doubt play a part in the future development of methods of culturing the species in captivity.

Chinook salmon (*Oncorhynchus tschawytscha*)

The common name of this species is taken from the Chinook Indians, a tribe on the northwest coast of America whose survival was linked to these salmon returning to the Columbia River. Local names include 'spring' and 'king', and very large fish are called 'tyee' salmon. The natural range in western America extends from California to Alaska. Chinook have also been successfully introduced to rivers in the South Island of New Zealand, where they have formed truly anadromous stocks.

Recognition

They are deep, stoutly made fish with a thick caudal peduncle. Any very large Pacific salmon over 14kg in weight is almost certainly a chinook and it is only in the middle-size range that they can be confused with coho. They can, however, be easily distinguished internally because they have many more pyloric caeca (140–185).

The backs of the male fish are often very dark and almost black. The sides above the lateral line and the dorsal and caudal fins are heavily marked with fairly large, black spots. The males usually become darker

43

and go reddish round the fins and on the belly while finding the way back from their ocean feeding grounds, before they have reached their parent rivers. The male spawning dress is spectacular. The body becomes dark red with a bright red tail fin. A heavy kype develops and the teeth lengthen into fangs. When the fish of both sexes are ripe to spawn, the skin becomes spongy and grows over the scale pockets.

The flesh is pinkish-red but coarser than that of coho. River-caught fish make excellent smokers if they are not too close to spawning.

Spawning   The local name 'spring' salmon originally derived from the early run which enters the Columbia River in April and May. This is followed by other distinct runs, a summer run in June and July and an autumn run in August and September. In some other rivers, the main run of chinook takes place in the spring but some fish continue to enter through the summer and on into the late autumn and winter if the rivers remain ice-free. The early entering fish usually travel furthest upstream to their spawning grounds and the late fish spawn nearer the tideways.

Adult chinook remain in fresh water for some time as they are inhibited from spawning until the water temperature falls below 12°C. Female chinook shed between 3,000 and 12,000 eggs depending on their size. The eggs are 6–7mm in diameter.

River life   Chinook can descend to salt water soon after hatching but a proportion remain in their parent rivers for a parr life of one or two years. They resemble coho and can only be told apart with certainty by the much larger number of pyloric caeca.

Marine feeding and   Chinook are piscivorous from an early age, feeding on
growth in the sea   herring and other fatty, shoal fish. The majority of young chinook which descend to the sea before they have completed a year of freshwater life return to

44

spawn after three sea winters. In some rivers five- and six-year-olds are common and the very large fish have usually had more than five years of sea feeding. In the most northerly rivers all the young chinook have a parr life of more than one year and most of the female fish returning to spawn are six or seven years old, with the male fish generally a year younger.

Chinook are the largest of the Pacific salmon. The average weight at the end of marine life is about 10kg, taken throughout the range of the species, but the largest fish can be over 45kg in weight and more than 150cm in length.

Fish farming   Chinook are being farmed commercially in net cages in the sea in western North America but they have not so far been of interest to fish farmers in Europe. The growth rate in wild fish is not as fast as in some other salmonids or other species of Pacific salmon but the fact that some races of chinook appear to mature very late and grow to large size could prove to be an attraction.

3 Atlantic salmon

The marine population of this species is made up of stock from both sides of the North Atlantic. The fish spawn in rivers on the mainland of North America and the larger off-shore islands from the Koksoak river in Ungava Bay to the St Croix which forms the eastern boundary between Canada and the United States. Atlantic salmon have also been re-established in some rivers in the state of Maine. There is a small population in two rivers in southeast Greenland and salmon enter and spawn in all suitable rivers in Iceland. In the British Isles and on the continent of Europe this species still enters the faster flowing rivers which remain unpolluted and unobstructed from the White Sea to the north coast of Spain, including rivers flowing into the Baltic. Genetic differences between the salmon spawning in rivers in eastern North America and in Britain were demonstrated in the early 1970s. Since that time genetic differences have been shown to exist between races in different rivers in the same country of origin.

Non-migratory populations of *Salmo salar*, which remain throughout their life cycle in fresh water, occur in the USSR, Finland, Sweden and Norway. The Norwegian freshwater race is a dwarf, relict type, but

the forms resident in the lakes in south Sweden and in
Lake Ladoga in Russia grow at rates comparable to
those of marine migrants. Freshwater populations also
occur in North America. The form resident mainly in
lakes is known as sebago salmon. They grow to a
weight of 6–8kg. Another form occurs in some rivers
on the north shore of the St Lawrence and in northern
Quebec Province. These fish are known by the Indian
name of *uananiche*. They are a good deal smaller than
sebago salmon and do not reach weights of more than
1–2kg. An interesting fish known as the Adriatic
salmon occurs in rivers in Dalmatia. It has been given
the status of a separate species and the name of
Salmothymus obtusirostris although it is a relict from the
last ice age and closely related to both Atlantic salmon
and sea trout (*Salmo trutta*).

Differences in the climate and geology affecting rivers
throughout the range of Atlantic salmon cause wide
differences in growth rates in fresh water and the time
of first migration to the sea. Relatively little is known
of the marine life of this species but fish from both
sides of the ocean meet on their sea feeding grounds.
Salmon from the Norwegian rivers are known to feed
mainly in the Norwegian and Barents Seas. Stocks
from the Swedish, Finnish and Russian rivers draining
into the Baltic Sea generally remain entirely within its
confines throughout their sea life.

The young fish during the period they spend in fresh
water are known as parr. The name changes to smolt
when they are ready to migrate to the sea. This name
stems from the silvery coating which develops on their
scales. The fish are seldom seen or captured during the
early part of their marine life. At this stage they are
known simply as post-smolts. They are silver coloured
without visible spots and very slim with a narrow,
deeply-forked tail. Although they become deeper and
fatter, with some visible black spots on their sides,
they retain much of this basic shape for the first or
'grilse' year at sea. The fish which return to their

47

parent rivers as grilse are still recognizable by their streamlined shape and the slim 'wrist' between the body and tail-fin which is appreciably more forked than in older salmon.

The fish of both sexes are bright silver when feeding in the sea and it is sometimes difficult to distinguish males from females. The flesh is red and contains a high proportion of fat. Salmon returning on spawning migration have developing gonads which reduces both the red colour and the fat in their tissues. This change is rather less noticeable in the rivers in the north of Norway which are relatively close to the sea feeding grounds of their native salmon. The flesh of farmed salmon which are fed up to a short time before slaughter on a wet food mixture with added fish oils and sources of carotene has much the same colour and content as that of wild fish caught when feeding in the sea, rather than when they are returning to fresh water on spawning migration. The external appearance of the fish, particularly the males, changes radically as the gonads develop. The lower jaw grows to form a hooked 'kype' on the male fish.

The adult fish must spawn in fresh water. At the onset of sexual maturity salmon return to their home rivers and if possible to the area where they hatched and spent their initial freshwater parr life. Atlantic salmon generally spawn in the autumn and winter in the period from October to January, although in some rivers there are stocks which spawn in February and March. Spawning time varies according to geographic distribution. The fish which spawn first inhabit those parts of the range of the species where the rivers freeze over early in the winter. The spawning period goes on longer and starts later at the southern end of the range in Europe where most salmon rivers remain ice-free in winter.

The female fish first selects a site where the gravel is of the right size and of sufficient depth, and where there is an appreciable current of water passing

48

through as well as over the stones in the river bed. She then excavates a hole by turning on her side and flexing her body up and down. In this way she can raise and move quite large stones which weigh comparatively little in water. Her body does not touch the stones. They are lifted through the suction created by her rapid movements. She continues the excavation until it has reached about the same length and depth as her body. The male fish takes no part in making the nest or redd but often fights with other cock fish that approach the redds.

When the female fish is ready to spawn, the male moves alongside the female and fertilizes the eggs as they are extruded. Sometimes a precocious male parr will slip in and successfully fertilize the eggs ahead of an adult male. The female does not extrude all her eggs at one time. After a rest period she moves a short distance upstream and excavates another trench. The clean gravel from the second trench is washed down and covers the eggs deposited during the first shedding. The process is repeated until the eggs have been extruded. In the final excavation virtually no eggs are left to shed and the gravel serves to fill in the previous trench. A pair of spawning salmon may spend a week or more on the redds, depending on the water temperature. Usually spawning is completed in two to three days. Female Atlantic salmon produce between 1,200 and 2,000 eggs per kilogram of body weight. Large fish produce more eggs per kilogram than small fish.

The length of the incubation period for salmon eggs depends on water temperature. When the incubation period has elapsed (approximately 440 day/degrees), the alevins hatch out. Their mouths have not developed and they continue to derive nourishment from the yolk sacs protruding from the ventral surfaces of their bodies. The yolk sacs are gradually absorbed over the next three to four weeks, again depending on water temperature.

49

The gravel plays a particularly important part in creating the right environment for both incubating eggs and developing alevins. The female fish has deliberately selected a site where there is a current passing through the stones. The chinks between the stones provide individual, quiet resting places for the eggs and subsequently for the alevins. The larval fish do not have to expend any energy in order to maintain their positions, which is a vital factor in their survival at a later stage.

The final absorption of the yolk sac coincides with the development of the mouth, digestive tract and excretory organs and the fry are ready to feed. They work their way up through the gravel and take station in a hollow between stones on the river bed. A dye test will show that there is a vortex in these hollows which prevents the fish from being washed away and allows them to remain in position with the minimum of effort. The fry feed on zooplankton drifted down to them in the current. As they grow bigger they gradually make longer forays after larger food organisms.

Salmon parr spend most of their freshwater life in shallow riffles, were the water is broken and well-oxygenated but the current is not strong. The time taken in fresh water to reach the size for migration to the sea depends on the length of the summer feeding period. At the southern end of their range a fairly high proportion of Atlantic salmon parr reach a length of 12–15cm, transform into smolt and are ready to migrate to the sea in the spring of the first year after hatching when they are 1 + years old. At the other extreme, in rivers in northern Canada and Arctic Norway, they may take five or six years to reach the smolt stage. The average age at smolting over most of the range of Atlantic salmon is 3 + years. The average in the British Isles is 2 +, but most rivers have a proportion of fish which become smolt after 1 + or 3 + years.

No one knows how long Atlantic salmon smolt spend

in home coastal waters before putting out to the open sea, and very little is known of the way they take or what directs them to take that way. The salmon, which enter the sea as smolt, feed on a changing diet as they grow in size during their marine life and it is most likely that movements to different areas in the sea depend upon the presence of the right prey.

At each stage of their marine life different races of Atlantic salmon probably go where there is an abundance of the food animals on which they are genetically conditioned to feed. Initially, post-smolt feed on amphipods (small, shrimp-like crustaceans), although very little is known of the feeding habits of the migrants going to the open seas as opposed to the Baltic. Pre-grilse have been caught when feeding on euphausiids (krill). At a later stage, salmon certainly become piscivorous and feed mainly on small fish. Favourite food species are capelin (*Mallotus villosus*), sandeels (*Ammodytes* spp.) and members of the herring family, which are all fatty fish forming dense, pelagic shoals. The diet also includes some bigger crustacea such as the Arctic prawn (*Pandalus borealis*). Arctic squid are also taken in some sea areas. One thing is certain, Atlantic salmon feed in the northern waters, a fact of the greatest importance to fish farmers.

Sea feeding areas used by Atlantic salmon are known to extend along the edge of the Arctic pack-ice and into the Labrador sea. Salmon from Norwegian rivers feed in the Barents and Norwegian seas between Jan Mayer Island, Svalbard and Novaya Zemlya. The surface waters in these sea areas remain cool in spring and summer, ranging from 2°C to 6°C and not rising above 8°C to 9°C. There is a definite indication that the races of Atlantic salmon feeding in the open seas prefer colder waters and that their growth rate is slower when the water warms up.

Atlantic salmon stocks in the wild are delicately balanced and are subject to over-fishing, mainly by drift netting on their marine feeding grounds or at sea while on spawning migration. Survival of the species depends on the preservation of their freshwater environment. Stocks are extremely sensitive to the effects of pollution, obstruction and other man-made changes in the natural regime of their parent rivers. Wild stocks are generally declining throughout the range of the species, which emphasizes the need to accelerate domestication before it is too late. Serological studies are now providing good reasons for believing that each salmon river may have its own distinct stock of salmon with genetically stable

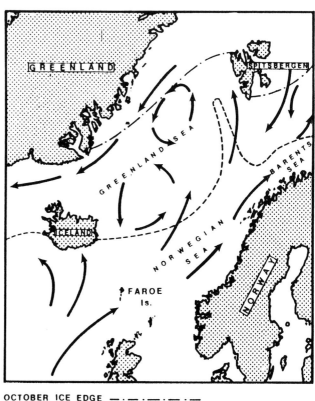

Fig 4 Ocean currents on salmon feeding grounds in the north Atlantic

OCTOBER ICE EDGE —·—·—·—·—
LIMIT OF DRIFT ICE ——————————
DIRECTION OF SURFACE CURRENTS ➝ ➝ ➝

52

characteristics. This is of great importance to the salmon farming industry because the wild stocks chosen for breeding in captivity can now be selected for attributes which make them most suitable for domestication.

Recognition There are wide distinctions between the external characteristics of races of Atlantic salmon but the scale arrangement and appearance is typical of the species. There are 10–15 (usually 11–13) scales counted in a forward-sloping direction from the adipose fin to the lateral line. The scales are cycloid and show growth rings, the relative position of which can be used to interpret the life history of the fish. They show the length of time it has spent initially in fresh water and the time spent feeding in the sea. Scales also show the occasions when the fish has spawned.

There are 10–12 rays in the dorsal fin.

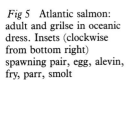

Fig 5 Atlantic salmon: adult and grilse in oceanic dress. Insets (clockwise from bottom right) spawning pair, egg, alevin, fry, parr, smolt

53

The head of the vomer (the bone in the roof of the mouth) has no teeth. The teeth on the shaft of the bone can be shed and replaced.

A recognition feature commonly used, particularly to distinguish salmon parr from small brown or German trout (*Salmo trutta*), is that the posterior end of the maxilla (upper jaw-bone) does not extend back beyond the eye. This can be a useful guide but is also sometimes confusing.

Hybrids can occur in the wild between European sea trout (the sea-going race of *Salmo trutta*) and Atlantic salmon. These fish may have a mixture of the distinguishing features of both species.

Growth, size and age at sexual maturity

Larval growth in freshwater is slow. The food supply may be limited but the young fish make a better use of the available food animals than other young non-migratory salmonids of the same age sharing the same environment. Larger food animals are taken as the fish descend to brackish estuarial waters and enter the sea, and marine growth becomes progressively more rapid.

A variable proportion become sexually mature and return to fresh water on spawning migration as 'grilse' (*ie* after one winter in the sea) when they have reached a length of 50–65cm and a weight of 1·5 to 3·5kg. The fish which remain a second winter in the sea and return to fresh water in the following year when they are approximately two years old are 'salmon' and are normally between 70cm and 90cm in length and 4–6kg in weight. A relatively small proportion of fish (greater from some rivers than from others) remain at sea for three years or longer. These fish become large salmon and after three years feeding at sea reach lengths of 90–105cm and weights of 8–14kg or more.

The longer Atlantic salmon remain at sea before returning to fresh water on their first spawning migration, the larger they become, but the oldest 'maiden' fish seldom exceeds four sea winters in age.

54

Fig 6 Return migration routes of Atlantic salmon from major feeding grounds

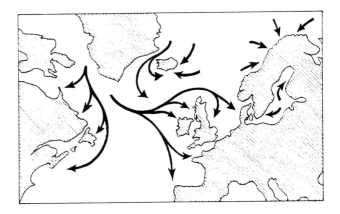

The males are generally considerably larger than the females. The largest Atlantic salmon taken by nets in the Tana River, which is in Finnmark in the far north of Norway and runs into the Barents Sea, exceeded 45kg (100lb) in weight.

Atlantic salmon appear (from scale reading) to feed continuously throughout their marine life, but more intensively in the spring and summer than during the winter. Indications are that the winter reduction in growth rate is less in the open sea populations than for those in the Baltic, where growth in winter is a good deal slower than during the summer, probably due to surface ice formation and lower water temperatures. On the other hand, growth in summer appears to be proportionately greater in the Baltic salmon. Salmon in water of full oceanic salinity (33–34‰) appear to continue to feed intensively at lower temperatures than populations in brackish water (and most probably populations in fresh water).

Comparable rates of growth for wild salmon, deduced from scale samples taken from fish caught on their sea feeding grounds in various months of the year, show that the rate of growth of Baltic salmon is generally slightly more rapid than that of salmon in the northwest Atlantic, but neither group grows as quickly as salmon feeding in the seas off the north coast of

Norway. These observed patterns of marine growth are probably much more indicative of genetic, racial differences than of the physical characteristics of the marine environment of the fish or the abundance and variety of the species preyed upon as food in the different sea areas.

The growth rates of freshwater races of Atlantic salmon in large lakes in northern Europe and North America, where shoals of fish species are present, are quite comparable to those attained in the sea. Atlantic salmon held experimentally in fresh water beyond the time of smoltification in Sweden and in Canada grew at a rate equal to fish of same year class in water of full marine salinity. The fact that wild populations of Atlantic salmon can grow as rapidly in a suitable freshwater environment as they can in the sea may have far reaching implications for the future of salmon farming.

4 Farming in the sea

The best salmon sea farmers are fishermen. They know and respect the sea. Working while balanced on a moving deck, has been their daily life. They are ready for sudden changes in wind and weather and are able to handle nets and gear in rough, bitterly cold water. Their understanding of fish as the source of their livelihood makes a base for the changeover from hunting to farming the sea.

The vast, three dimensional space of the seas and oceans now need to be explored to find good sites for sea farms. It is the way the waters move, together with their chemistry and temperature, which must first decide which places are suitable for development.

Seas and oceans   Oceanic currents in both hemispheres flow from the warm water near the equator round the ocean basins towards the poles. The rotation of the earth directs the currents clockwise in the northern and anticlockwise in the southern hemispheres. In the north, where the warm currents meet the cold Arctic water are the rich feeding grounds of the Atlantic and Pacific salmon.

In the northwest Atlantic there is no temperate zone in the sea and in winter the eastern coast of the North American continent as far south as Cape Hatteras is in

57

the icy grip of the Labrador current. Ships are coated in ice in the harbour of Halifax, Nova Scotia, which is on the same latitude as the South of France. The Gulf Stream warms the eastern Atlantic as far north as the Norwegian Sea where the surface water temperature seldom falls below 4°C. The same pattern of sea temperature and climate is repeated in the Pacific, where the warm ocean currents moving northwards flow away from from the coasts of Japan and northeast Asia and cross over to the shores of British Columbia, which has a climate similar to the coastal areas of Europe which are warmed by the Gulf Stream.

Coastal currents   Where large rivers run into the sea, currents of lighter, less saline water form and flow along the coast inshore and usually in the opposite direction to the ocean currents. Coastal currents in the northern hemisphere flow generally north in Europe and on the western seaboard of North America, and south along the American Atlantic coast and the northern Pacific coast of Asia. The boundary between oceanic and coastal currents is indicated by changes in the temperature and salinity of the water.

The intensity of coastal currents varies with the seasons. They are of importance to fish farmers not only because of the flow of water through a farm site

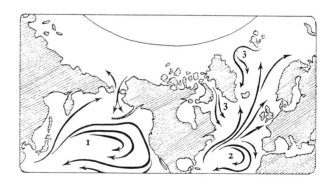

Fig 7 Ocean currents of the northern hemisphere: 1 Japan Current; 2 Gulf Stream; 3 Arctic Current

but also because of the changes they may produce in salinity and water temperatures. These may be detrimental to fish held in cages or promote dangerous growths of poisonous or suffocating plankton.

Tides and tidal currents   The moon takes 24 hours and 52 minutes to rotate round the earth and its passage is followed by two tides which become later by 52 minutes each day. The sun also causes two much weaker tides during each rotation of the earth. At full moon, the sun and the moon are on opposite sides of the earth and at the new moon they are both on the same side. Their gravitational forces then act together and produce the extra high and extra low 'spring' tides. At half moon, the sun and moon are at right angles and counteract each other's gravitational pull. This results in the small 'neap' tides. The succession of 'spring' and 'neap' tides progresses through the lunar months of 29 days.

A continuous tidal wave cannot follow round the earth's surface in the wake of the moon because of the intervening land masses. The oceanic tides flow round the ocean basins leaving central zones with very little tidal rise and fall and causing progressively greater tides on the more distant shores. The mechanics of the tides may be of academic interest but tidal movements in coastal waters are a vital factor in the selection of sites for sea farming. The deep ocean tides cause little lateral movement, but in shallower areas the water is shifted by tidal currents and changed at least twice daily in bays and inlets. Without this movement and interchange, cage sites would become low in oxygen and fouled with waste products.

Salinity   The water of the open oceans and seas holds between 3 and 3·5% of salts. Salt content is usually measured in parts per thousand and oceanic water has a normal salinity of 30–35‰. Only the concentration varies from brackish such as in parts of the Baltic Sea to the

59

strongly saline water of the Mediterranean. The
principle salts in sea water are:

	mg/l
Sodium chloride	27·13
Magnesium chloride	3·85
Magnesium sulphate	1·66
Calcium sulphate	1·26
Potassium sulphate	0·86
Calcium carbonate	0·12
Magnesium bromide	0·074
Trace elements	0·0035

Alkalinity   All marine animals are highly sensitive to changes in
alkalinity. The carbon dioxide that enters the water
from the air, or is produced by the respiration of
marine animals and photosynthesis by plants, forms
carbonic acid and bicarbonates in sea water which have
a buffering action. This prevents rapid changes in the
carbonate equilibrium and maintains a stable pH, and
is one of the benefits gained by ongrowing salmonids in
the sea, provided the salinity remains constant in cages
or enclosures.

Aeration   The life of fish in the sea and in large lakes depends
mainly upon the mixing of air and water at the surface
by the wind and waves. In summer, during periods of
calm weather, plants in the depths reached by sunlight
also make a significant contribution to the oxygen
content of the water.

Temperature   The water temperature of the surface layer in
temperate climates and in the north where coastal
waters are warmed by ocean currents seldom drops
below 3–4°C. In summer the water temperatures do
not rise much above 10–15°C. By contrast, surface
water in the Arctic and Antarctic can remain close to
1°C all the year round and summer temperatures rise
to 30°C in the Red Sea.

60

Freezing   As ice forms, the frozen water loses salt and the salinity below the ice increases. When the ice melts, the fresh water floats over the surface layers and salinity falls rapidly. Ice formation on the surface of waters used for cage culture is not necessarily fatal to the fish but when the ice breaks up, wind-driven ice-floes can wreck fish cages or enclosures.

The freezing point (T) of sea water depends on its salinity and can be found by the equation:

$$T = -0.054 \times S \text{ (where S = the salt content in parts per thousand).}$$

For example, brackish water with a salinity of 10‰ would freeze at $-0 \cdot 54°C$, but sea water with full oceanic salinity of 35‰ freezes at $-1 \cdot 89°C$ and water with a salinity of 20‰ would freeze at $-1 \cdot 1°C$. Very cold water induces increasing passivity in salmonids and it is reckoned that they can only withstand temperatures down to about $-0 \cdot 5°C$. It is therefore possible to get critically low temperatures which could be lethal without ice being formed. In some northern waters, where in winter the sea water temperature is maintained above freezing point by warm ocean currents, snow in blizzards over rough seas can mix with surface water causing temporary super-cooling. This can be fatal to fish held in cages where they cannot escape to deeper water.

Pollution   The blueness of the sea is not an indication that it is either clear or clean. Ocean waters are a mirror for the sky although they can contain some colouring matter. Inshore waters can be coloured by plants or animals and by detritus washed down by rivers or disturbed from the sea bed in stormy weather.

Pollution of the sea in coastal areas is becoming an increasing problem and can limit the choice of sites for sea farming. The effect of man-made pollution can be either directly poisonous to fish life or can change the water chemistry in such a way as to encourage the

61

growth of poisonous or suffocating plankton. The risk of oil pollution is now an ever present menace to the sea farmer.

Location of sea farms   Some basic hydrographic features are common to all cooler waters in the northern hemisphere that could prove suitable for salmon farming. Atlantic oceanic water is distinguished by a stable salinity in the region of 35‰ and a small, stable temperature range. Coastal waters in areas where they are influenced by the influx of fresh water or the intrusion of Arctic currents usually have a variable and generally lower salinity and a more variable temperature range than the open ocean. A high salinity generally indicates stable salinity while low salinity is usually variable.

Conditions in coastal water are governed by the overall behaviour of currents and the average movement of inshore water masses resulting from meteorological factors and the effect of tides. Variation can be short-term over a few months, annual or medium-term, and long-term, taking place slowly over a period of years. Short-term changes are indicated by sudden local alterations in water temperature and salinity, and in the speed and direction of currents. They result from the effects of radiation, evaporation or the rapid influx of fresh water. Such changes are often brought about by an alteration in local tides caused by gale-force winds.

In the northern hemisphere surface water is pushed off shore at right angles to the wind direction. This means that on coastlines running north and south a northerly wind forces surface water away from the shore which is then replaced by colder water from greater depth. A prolonged period of strong northerly winds during the summer months can lead to a rapid drop in coastal water temperature of as much as 5–7°C. Southerly winds tend to produce the opposite result. The general effect of wind on coastal water conditions can be modified to a great extent by the geography of a particular area and local water temperatures may not conform to the expected pattern. A discrepancy is most likely in the confined waters of fjords and sea lochs, particularly where there is a substantial inflow of fresh water.

Some fjords and most sea lochs have a well-defined threshold between open sea and the deeper water on the landward side. This can result in the strongly saline water from outside the threshold forming a steeply sloping boundary surface with the over-lying, less saline water in the sea loch. Interchange of water between the sea loch and the open sea can be prevented, particularly during the summer months, by the formation of a boundary surface at a threshold, and stale water of lower oxygen content and salinity then circulates back and forth inside the enclosed area. The effect of the formation of a boundary surface is likely

Fig 8 Salmon on-growing enclosure between islands in the sea. The fish feeding pipe extends along the centre gangway

to be less noticeable in places where the tidal range is greatest.

There is a wide variation in the average ranges of tides on those parts of the European and North American coasts suitable for farming salmonids in cages or enclosures. An indented coastline with off-shore islands can result in large local differences. Where there are eddies and confused currents, neutral zones can occur with practically no tidal flow. The strongest flow in sea lochs usually occurs at half tide.

Site exploration   The essential factors governing the selection of a sea site can be summarized as follows:

— prevailing wind and weather;
— local geography;
— exposure;
— water depth;
— bottom formation;
— tidal flows and currents;
— water temperature, salinity and chemistry.

The initial sources of information are maps and charts. All relevant reference books should be consulted. Marine sailing directions can give a good deal of information on the local geography and hydrography of inshore waters. The broad characteristics of weather systems may be of less importance in the selection of sea-site location than local modifications and the protection afforded by nearby land. Exposure to wind and wave action is a basic factor for consideration in site selection. Maximum 'fetch' (distance to the nearest land) should not exceed approximately 3km in the direction of the prevailing wind. A longer fetch can sometimes be tolerated in otherwise well-sheltered waters but a site should remain workable in any force of wind from any direction.

When a sea area has been found which is sufficiently sheltered to provide a safe anchorage for cages, it must

64

be surveyed to establish the depth and bottom formation. There should be a depth of at least 10 metres below the cages at neap tides. This is essential to avoid the risk of accumulated excreta and waste food contaminating the water in which the cages are suspended or promoting eutrophication and abnormal plankton growth.

A bottom survey should include a wide enough area to cover the entrance to any bay or sound in order to make sure that a threshold will not interfere with the interchange of water from the open sea. The type of bottom is a useful indication of deep currents. A muddy bottom usually indicates a poor interchange of water at depth. The converse is indicated by a clean bottom. An absence of bottom fauna, such as crustacea, often indicates a stagnant area and should be avoided. Sites in channels with sufficient depth, close to large shallow, muddy or sandy areas should be avoided as the water may become turbid in winter gales.

Tidal flows and currents Currents close to the surface are essential in order to bring clean, well-oxygenated water to suspended cages or enclosures. In shallower water, currents near to the bottom may be needed to prevent accumulation of waste products, otherwise the cages may soon have to be moved to a new anchorage. The average current over a site should be greater than 100mm per second and not more than 500mm per second, otherwise net cages may be distorted and damaged. Beware of sites between islands or off peninsulas and avoid deep channels away from rocky areas as there are likely to be strong, scouring currents and cages cannot be safely anchored.

Salmon in sea cages suspended in clean, well-oxygenated water, will tolerate a nil flow for about one hour over dead low water, but the fish should not be fed during this period when sea temperatures are high during the summer months. The movement of fish held at fairly high density in small cages will cause a direct interchange of water which is sufficient to keep them alive, provided the cages are not too close together.

Temperature   Little is known of the most favourable temperature for growing salmon in cages in the sea. In the wild, the water temperature over their feeding areas is likely to be in the region of $8-10^{\circ}$C during the summer. The optimum temperature for on-growing some members of the salmon family is known from experience to be $15-16^{\circ}$C. Pacific salmon will tolerate comparatively high seawater temperatures and appear to do reasonably well in water up to 15°C.

The temperature range is greatest in confined sea lochs where the salinity is relatively low. The surface water in an enclosed area with a narrow opening may have a considerably higher temperature in summer and lower temperature in winter than more open waters, even if the salinity remains relatively high and stable.

Salinity   It is evident that the salinity of the environment has some direct effect on the growth rate of the sea-going races of salmonids and that racial characteristics are the dominant factor in saltwater tolerance. Races or species which migrate to the open sea and spend their period of most rapid growth in water where the salinity is 33–35‰ would be expected to make optimum growth if kept under similar conditions in captivity. Fully anadromous species or races appear to be stressed by changes in salinity. Very little is known on this subject and research so far carried out has not provided any precise information.

Oxygen   The oxygen content of sea water varies over the year and is at a maximum in early summer and a minimum in early winter. The amount of oxygen in sea water depends upon the water temperature, the salinity and

66

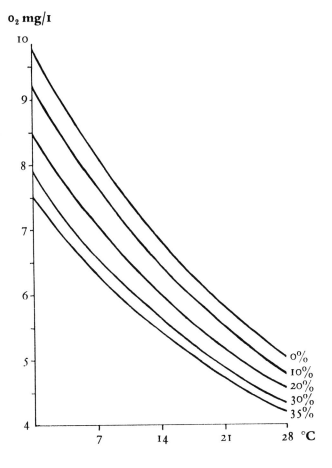

Fig 9 Dissolved oxygen in sea water at different water temperatures and salinities

atmospheric pressure. For practical purposes, at an atmospheric pressure of 760mm of mercury the amount of oxygen can be determined by the following equation (Truesdale and Gameson 1956):

$$\text{Oxygen in mg}/l = \frac{475 - (2 \cdot 83 - 0 \cdot 011 \text{ T}) \times \text{S}}{1 \cdot 38 \times (33 \cdot 5 + \text{T})}$$

Where T = temperature and S = salinity in parts per thousand.

The amount of oxygen that fish need depends mainly

67

on water temperature and their degree of activity. The amount needed increases when fish are feeding and rather unexpectedly decreases in proportion to their size. This means that the density at which the salmon can be kept in sea cages increases as they become larger. The presence of free carbon dioxide in the water is also known to increase the demand for oxygen in fish.

Acidity and alkalinity   The pH of sea water in northern coastal waters, where photosynthesis is taking place, is generally between 7·5 and 8·5. The pH tends to rise in summer when photosynthesis is at a maximum and to fall in winter. The pH in a sea-cage site area should not exceed 9·0 or fall below 5·0. There is little risk of the pH rising to a lethal level but the decay of waste products accumulated below cages and in enclosures, or held in the site area by a threshold at the outlet to a fjord or sea loch, could cause the pH to fall to a dangerously low level.

Water quality   The main risk of pollution is from the accumulation of waste products from the fish themselves. It is obvious

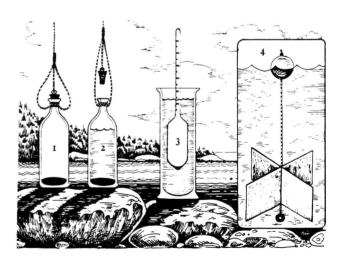

Fig 10 Exploration equipment: 1 and 2 sampling bottles; 3 hydrometer; 4 current measuring drogue

68

that a site should not be chosen where there is a risk of pollution from outside sources, either as a result of human activity or the natural decay of organic matter. The danger following the breakdown of accumulated faeces and unconsumed food is from free ammonia being released at a level that becomes actively poisonous to the fish which are held in a cage or enclosure and cannot escape to clean waters. It is as well never to overlook the fact that an otherwise attractive site for a sea farm can become a self-polluting death-trap if there is an insufficient depth of water below the cages or no current at the sea bed to disperse rotting waste matter collecting on the bottom.

Surveying a site
Measuring wind and wave forces

An anemometer is the accurate way of measuring wind speeds but time may not permit sufficiently long-term recording for the results to be of any real use. A guide to wind force over a period can be obtained from 'tatter flags' which are pieces of material fixed to a staff that shred at given wind speeds. Wave formation and the height and length of the sea are more important than wind strength. These are a consequence of 'fetch' as well as wind force and the relationship between fetch, wind speed and wave height can be plotted.

The Meteorological Office can be a useful source of information but inshore fishermen or other people who use the sea in the general area of the proposed sea-farm site are the best sources of information. A safe anchorage for small boats is a safe anchorage for sea cages, as far as wind and wave forces are concerned, but is usually too shallow.

Depth and bottom formation

The general depth of water and some idea of the formation of the sea bed can be read from the chart. A more detailed, local survey, using a depth meter should be made and checked by sounding. The appearance of the shoreline will give some indication of what the bed is likely to be and this can be checked from local sources of information. The best way to examine the sea bed in shallower water is by skin-diving.

Tidal flow measurement   There are various ways of measuring the speed of tidal currents. A simple flow-measuring drogue consists of a buoy on the surface which supports cross-shaped plates moved along with the current. Its speed of travel can be measured over a given distance from an anchored boat.

Net cages can be weighted to prevent their being pulled out of shape in stronger currents but the angle of the net to the perpendicular should not exceed 15 degrees or there will be damage to gear. The angle likely to be made by a net in a current can be roughly checked using a drogue adjusted to hang at a depth equal to the bottom of the net cage and suspended from an anchored boat.

Measuring temperature   The water temperature at the surface is easy to take
and salinity   but the thermometer must be given time to stabilize and should preferably be read in the water. The specific gravity of sea water depends upon temperature and salinity. The salinity at the surface can be measured by taking the temperature of the water and using a hydrometer to measure the specific gravity. Samples of water below the surface can be taken using a weighted flask sunk to the required depth. A cord is jerked to pull out the cork. The flask is then given time to fill and drawn up. The water in the neck is thrown out and a thermometer put in to take the temperature. The sample is poured into a measuring cylinder and the specific gravity read off with a hydrometer.

In practice, it is seldom necessary to measure salinity with an accuracy of more than ± 2‰. The difference in specific gravity produced by changes of ± 5°C on an average sea water temperature of 10°C is so small that for practical purposes it can be disregarded and the salinity can be estimated from a hydrometer which can be directly graduated for salinity.

Sites for enclosures Siting any type of enclosure is governed by the tides. All enclosures require the erection of a fish-proof barrier from the surface to the bed of the sea which has to withstand wind and wave action. It is impracticable and uneconomic to put up a barrier high enough to operate over a large range in tidal rise and fall. In practice the difference in water level between high and low tide should not exceed about 1 metre at ordinary 'springs'. Although the tidal ebb and flow must be small, there must still be sufficient flow to change the water inside the enclosure and to carry away waste.

The tidal current through large enclosures made between islands or in open-ended channels may prove insufficient and have to be supplemented by pumping. This method can still be profitable for a large-scale development. It is difficult to find a satisfactory site for an enclosure to be made by erecting a barrier across the seaward end of the inlet as there as to be a good tidal current through the site if fish are to be kept at a worthwhile density. Both these types of enclosure need

Fig 11 Salmon cages in
Arctic Norway

71

a clean bottom, free from rocky outcrops and from places of attachment for weeds.

Easy access is vital by both water and land. There must be a satisfactory area for a landing stage with an access road from which cages and enclosures can be serviced and where fish can be landed and equipment brought ashore for repairs. The nearer the landing stage is to shore-based stores and fish processing premises the better. The cost in time of transporting staff, fish food and fish over unnecessarily long distances, either by sea or land, can make any sea farm unprofitable to operate.

No matter how much hydrographic or other investigation is carried out in advance of selecting and commencing operations on a sea farm, some of the potentials of the site, good and bad, will only be properly appreciated when it is being worked and begins to rear fish. Sea cages are generally portable and can be moved to another site. It is therefore prudent not to make any large investment in either fixed marine or inshore facilities until a pilot project has been carried out and the results have been assessed.

5 Brood stock and egg production

Salmonid farming is still at the stage in which wild races of the different species are being tested to determine those suited to domestication. The 'homing' of anadromous salmonids has led to the development of separate populations with a gene pool corresponding to the environment to which they constantly return on spawning migration. The physical separation of different stocks resident in fresh water has also produced distinct racial characteristics. Blood-typing of salmon has demonstrated genetic differentiation. The second stage, which is yet to be generally explored, is genetic engineering designed to improve the quality of the selected, cultivated species.

The most sought-after features in salmon are as follows:

— Unstressed tolerance of an artificial environment.
— Short freshwater life before smoltification.
  Production of S1 smolts in Atlantic salmon.
— Resistance to disease.
— Ease of acclimatization to sea water.
— Good food conversion and rapid growth to market size.
— Delayed gonad development and delayed onset of

sexual maturity.

— Marketability. Flesh quality and colour. Body shape for processing.

The Norwegians are the leaders in the field of selection and the testing of various strains of Atlantic salmon. They have a very wide range of rivers to draw upon for wild stocks, most of which have distinct racial characteristics. Considerable variation in resistance to disease has been found as well as differences in marine growth rate.

Strains of Atlantic salmon from 14 different rivers in Norway and Sweden (Baltic Sea salmon) were hatched, reared separately and transferred to sea cages as S1 smolts. At a total age of three years, the average weight of fish belonging to stocks hailing originally from the Norwegian river Jordalsgrendea was 5·1kg. Strains from other rivers in Norway averaged between 4·1–4·9kg. The fish from the Swedish rivers did not grow as large as those from Norway and the strains which grew more slowly in the sea tended to produce a higher proportion of fish which matured as grilse.

Hybridization   The purpose of experimental attempts to create hybrids between the species of salmonids has been firstly to produce more useful characteristics and secondly to induce sterility. It is possible to cross-fertilize the eggs of most salmonids but survival during the alevin and fry stages is very variable and can be virtually nil. In the majority of cases it is too small to be of any practical use and many of the hybrids are only of

Table 1   Norwegian hybrid salmonids: percentage hatch of fertile eggs

Male		*Female*	%
Atlantic salmon	×	Arctic char	59
(*Salmo salar*)		(*Salvelinus alpinus*)	
Arctic char	×	Atlantic salmon	89

74

scientific interest and not at present of any value to fish farmers.

Atlantic salmon have been successfully crossed with brown trout and sea trout (*Salmo trutta*) but the hybrid does not seem to have any particular advantage over the parent species. The most successful hybridization so far achieved in Norway has been between Arctic char (*Salvelinus alpinus*) and Atlantic salmon (*Salmo salar*).

Hybrids have been produced between most of the Pacific species. Sockeye, pink and chum salmon have all been crossed successfully in either sex direction, with the exception of male pink × female sockeye which proved unsuccessful. Spring or chinook will cross with all the other North American species of Pacific salmon in either sex direction except female spring with male pink. In some cases the hybrids proved fertile.

Sex changes and sterilization
The physiological changes which take place as Pacific salmon approach sexual maturity render them totally unmarketable. Similar changes in maturing Atlantic salmon and the appearance of secondary sex characteristics greatly reduce their market value. Female Atlantic salmon mature later (older) than males and it is considered an advantage to produce and rear all female fish. Males on the other hand grow more quickly and could be a better proposition on a farm using very large cages where fish stocked as big smolt in April/May are harvested in November/December of the same year.

The method used to produce all female salmonids is to feed the fry a small amount of male hormone for the first month. All the fish then become 'male'. The sex-changed females grow testes and can produce fertile sperm. If the sperm from these fish is then used in the next generation to fertilize normal females, their progeny will be all female. A sperm bank from sex-changed females can be stored by cryopreservation

75

(cold storage in liquid nitrogen).

Triploid female salmon (fish which have three instead of the normal two sets of chromosomes) are sterile and do not develop mature sex organs, or (in the absence of the necessary hormone changes) secondary sex characteristics. Triploid males are useless as they are abnormally masculine producing all the undesirable, physiological changes at the onset of sexual maturity, but their seminal fluid contains no active spermatozoa. Various ways have been tried to induce triploidy in the progeny of second generation all female salmon. These are generally based on shocking the newly fertilized eggs. Methods used have included heat, pressure and chemical shock. Eventually a commercially satisfactory way will no doubt be found to induce triploidy in all female stocks of both Pacific species and Atlantic farm salmon.

Brood fish The selection and keeping of brood stock is not necessarily a job for specialists and all fish farmers should know something of the basic procedures. The initial step is simple selection. The best fish are graded out, which in practical terms means those that grow to largest size in the shortest time at each stage in fresh water and in the sea. A stockman soon learns to spot the good 'doers' in the animals he tends. These are the ones to select and keep back as brood stock.

Aspects of nutrition, hygiene and disease prevention are dealt with elsewhere. The holding arrangements for brood fish are not different to those for fish intended for market but they should be kept at a reduced density.

Female kelts of Atlantic salmon from farm stock survive well after spawning and large fish are worth 'mending' as they will produce good quality eggs on second maturation. Some males may also be worth retaining alive if they can be marketed but are seldom worth re-conditioning as brood stock. Pacific salmon of all species die after spawning for the first time.

76

Falling water temperature combined with increasing gonad development naturally reduces the fish's appetite. Salmon should cease to be given food when secondary sex characteristics become apparent and the males and females should be separated at this stage. If the fish are in sea cages ($200-300m^3$ or larger) they should be moved into smaller, rectangular cages. These should be $3-4m^2$ and 3m deep. Ropes should be attached to the nets and pass down under the cages and up on the opposite side. When these ropes are pulled up the bottom of the net cage can be raised to form two or four separate sections. The fish can then be easily checked for ripeness.

Brood stock on a shore-based unit can be kept in ordinary large tanks until they are approaching spawning time. They should then be moved into smaller tanks and the sexes separated. At this stage it is better to provide a freshwater supply, if sufficient is available on the site. Each tank should have a 'crush' which consists of a movable grid or wall of net that can be shifted to enclose the fish in a small part of the tank for ripeness testing. The tanks should be covered and the fish kept in semi-darkness.

Hatchery practice   Salmon can be successfully stripped of their eggs after being taken directly from the sea, provided they are washed in fresh water before and after being anaesthetized. The eggs can be fertilized on the spot and removed to the hatchery when they have completely 'hardened', or sperm and eggs can be transported separately. Sperm remains viable for several hours if kept cool in a vacuum flask and completely free of water. Unfertilized eggs also remain viable if they are in ovarian fluid and free of water. An alternative method is to bring the ripening fish ashore and store them in semi-darkened tanks until they are ready to be stripped. When ripening brood fish have to be stored in shore-based tanks for any length of time, they are healthier and easier to keep free from fungus

77

attack if the tanks have a supply of both fresh and salt water.

Anaesthetizing fish Before the widespread use of anaesthetics it was common practice to put some form of halter on large salmonids so that they could be more easily handled for stripping their eggs. Atlantic salmon were often suspended by a noose round the body under the pectoral fins while they were stripped.

A good, inexpensive fish anaesthetic is ethylaminobenzoate. A commercial product is marketed under the name of Benzocaine. A concentrated stock solution can be made up by dissolving 100g of the powder in 2·5 litres of acetone. A 25ppm working mixture is then produced by diluting 6·2ml of the concentrated stock solution with 10 litres of water. Increasing the strength of the mixture will reduce the time taken to immobilize the fish.

Fish taken directly from sea cages for stripping should be washed in fresh water before being immersed in the anaesthetic solution (remember that if

Table 2 Fish anaesthetics

Chemical	Dosage (ppm)	Immobilization time-min	Recovery time-min	Notes on use
Carbon dioxide solid acid-bicarbonate	100–200	1–2	—	Used for killing fish
Quinaldine 2-methy-quinoline	10–15	2–4	3–5	Non-critical Expensive
2-phenoxy-ethanol	40	2–5	5–10	More concentrated solutions up to 100ppm may be needed for large fish in warmer water Recovery time may be longer
Ethyl-aminobenzoate 98% in acetic acid	See instructions	2–5	5–10	A good, cheap anaesthetic

78

Fig 12 An awkward handful! Preparing to strip a salmon before the days of anaesthetics

the fish are dripping water each time one is put into the anaesthetic solution it will soon become weak and need renewing or topping up).

If they are intended to be kept alive and 'mended', the kelts should be returned to salt or brackish water as soon as they have recovered. Recovery should take place in a flow of clean, well-oxygenated water with the fish held upstream in the normal swimming position. It is possible to make a simple 'holder' for the fish during recovery. This consists of a rectangular trough divided into parallel channels along its length which are just wide enough to hold and support the fish. Baffles at the leading end divide the flow between the channels which are screened at the downstream end with taut netting.

Stripping eggs

There are no hard and fast rules about the fish-stripping operation, which must be learned by demonstration. Most fishmasters have their own techniques. The main essential is to have ripe male and female fish to hand in small tanks of convenient size where they can be easily caught up with a minimum of stress. Separate tanks will be needed for holding male fish, female fish, anaesthetizing the fish and for recovery. Various mechanical methods have been tried out for stripping eggs including air and water bags which compress the fish and squeeze out the eggs or sperm, but nothing is really as satisfactory as the human hand. Sperm intended for storage can be drawn up directly from the vent of ripe males using a pipette or a large syringe, which is easier than trying to run the sperm directly into a container. Fish of both sexes must be washed clean of anaesthetic solution and carefully dried with a cloth before they are stripped.

Fertilization

Salmon eggs are relatively easy to handle before, during and after fertilization. The more simple causes of loss are crushing while stripping, exposure to ultra-violet light in bright sunshine, and frost. Infertile eggs

79

Fig 13 Female Atlantic salmon ripe for stripping

Fig 14 Stripping eggs from a salmon

are not only a loss but are wasteful of space in the hatchery as they are difficult to detect before routine 'shocking' is carried out at the eyed stage.

Eggs and sperm carefully stripped from healthy, uninjured fish, when mixed together out of contact with water should normally result in 100% fertility. Eggs become very sensitive to disturbance for a period after water has been added and swelling starts. Losses are likely to be greatest due to movement during the first thirty minutes. Eggs take in water more quickly at higher temperatures. Swelling takes about $1\frac{1}{4}$ hours at $6 \cdot 5°C$ but only about 25 minutes at $13°C$ (spawning of some species of salmonids has been found to be unsuccessful at temperatures above $13°C$). The commonest cause of loss during the sensitive period following fertilization is from washing off the surplus sperm. If the eggs are washed (at $8°C$) between 4 and 15 minutes after water has been added, 40–50% may die over the next 48 hours. Losses of up to 20% can

Fig 15 Male Atlantic salmon big enough to fertilize the eggs from several females

occur if the eggs are disturbed for up to one hour after being placed in water.

Fertilized eggs can be quite safely kept without washing in a mixture of ovarian fluid and sperm for periods up to $1\frac{1}{2}$ hours then put directly into the water in the hatchery troughs. The alternative safe method is to wash the eggs at once, about one to two minutes after sperm is added, then leave the fertilized eggs standing completely undisturbed in a plentiful amount of clean water for ninety minutes before they are moved.

A practical method A simple stripping method is to anaesthetize a number of male and female fish (dependent on their size and the time it takes to strip one fish). When the fish are quiescent, wash them in clean running water. Dry each fish carefully with a cloth and put them into an insulated polystyrene box. Strip each female fish into a plastic bowl. When all the females are stripped, run in the sperm from two to three males and stir gently. The eggs can then be poured into a bucket and the process repeated with another group of male and female fish.

81

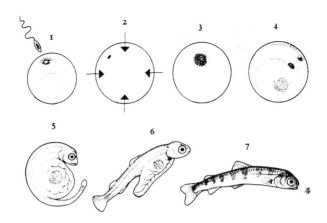

Fig 16 Development: green egg to fry. 1 Fertilization; 2 Egg swells; 3 Cell division starts; 4 Eyed stage; 5 Hatching; 6 Yolk-sac alevin; 7 Feeding fry

When the first batch of fertilized eggs have been in the mixture of sperm and ovarian fluid for about one hour, all the eggs that have been stripped and fertilized can be transferred to the hatchery troughs. Standing eggs must be kept cool and the bucket or egg container should be in an insulated box.

Each group of hatchery troughs should have its own litre measure. The eggs are poured directly into troughs from the measure. Eggs occupy less space before swelling and this must be taken into consideration; 1 litre of swelled or 'hardened' eggs is approximately equivalent to 1·4 litres of eggs before water is added.

Effect of light on eggs Exposure to light in the blue or violet part of the spectrum can be deadly to salmonid eggs either in air or water. Eggs should be kept in the dark. When artificial light has to be provided in a hatchery this should not be from fluorescent tubes. Yellow or orange light is safe to use but ultra-violet light (UVA) can penetrate to considerable depth in water.

Disinfection Imported eggs bought in during the eyed stage should always be disinfected. Recommended disinfectants are iodophors. These are solutions of iodine in organic solvents. The manufacturer's recommendations should be followed for trade preparations.

Buffodine is a buffered iodophor for the disinfection of eyed eggs (where the eyes of the embryo are visible through the chorion) and newly stripped, non-hardened eggs.

Cryopreservation of salmonid sperm (and eggs)

Considerable interest has been aroused during recent years in the possibility of creating and maintaining a sperm bank for farm salmonids. This would avoid having to keep male brood fish and might lead to the general use of sex reversal in both brood and market stock which would then be all-female. The problem has been to find a suitable 'extender' for fish spermatozoa. This is a chemical compatible with seminal plasma, such as a modified Cortland solution, that can be used to dilute the semen for long-term storage at very low temperatures. At the same time a compound such as dimethylsulphoxide (DMSO) has to be added to protect the sperm cells from damage during freezing and thawing.

Various methods have been tried out to obtain rapid freezing. A technique developed by Stein at the University of München has proved most successful and has resulted in a high percentage of fertilization using cryopreserved sperm. The diluted and prepared semen is allowed to fall in droplets on to 'dry ice' at $-78\cdot6°C$. The frozen pellets are then stored in liquid nitrogen (LN2).

Cryopreservation of sperm from Atlantic salmon and from four species of Pacific salmon—pink, coho, chum and sockeye—has been shown to yield actively motile spermatozoa capable of fertilizing a useful percentage of eggs. The commercial development of long-term sperm storage will probably follow stock improvement and the creation of particularly valuable strains. Attempts to cryopreserve unfertilized eggs of teleost fish has not shown much promise of success, probably due to the size and complexity of the cell structure.

Egg densities Atlantic salmon eggs should not be more than one layer deep in the hatchery baskets. The eggs are then clearly visible and much easier to pick and keep clean.

The available flow of water at 100% saturation with oxygen should be $5l$/min for each 10,000 eggs in the hatchery.

Incubation The standard incubation procedures for salmonid eggs are quite straightforward. Removal of dead eggs, which are easily recognized because they go white, is best carried out each day in smaller hatcheries producing fish for on-growing on the same unit. This can be done in various ways using a continuous siphon or a suction bulb. If not removed, the dead eggs will form a focus for fungus which will then attack surrounding healthy eggs. A continuous record of the dead eggs removed should be kept.

When the eggs under incubation have become eyed (the eyes of the embryo becoming visible as black spots through the egg shell or chorion) they should be 'shocked'. A simple way to do this is to lift the egg trays out of the water and leave them in the air for a few moments propped across the trough. The effect of 'shocking' is to show up infertile eggs, and any weak eggs, which will then go white and can be removed.

Specialist farmers using flask incubators, in which the eggs form a thick column in a vertical container with an upward flow of water, or hatcheries where very large quantities of eggs several layers thick are incubated in troughs, may have to use a fungicide bath to kill off fungus spores, without trying to remove the dead eggs.

Fungicides for use on fish or fish eggs The fungicide commonly used in the past was a compound known as malachite green. This substance is now alleged to be dangerous to the people who use it over long periods (the hands of most freshwater fish farmers are generally stained green). A new fungicide has now been developed for the treatment of eggs under incubation. This is called proflavine-

84

hemisulphate. The eggs are immersed in a solution of this compound at a strength of 1 part in 40,000 for one hour (the arms of fish farmers using this substance are now generally stained yellow up to the elbows).

Egg counting   Automatic egg counters, which also extract dead eggs, have been developed for use by specialist egg producers. The eggs are put through the machine when they have eyed-up. In one type, a disc with holes corresponding to the average size of eggs is fixed in position and then automatically loaded with eggs from a hopper as it revolves. Any dead eggs (which are opaque) intercept a beam of light as the disc, in which they are held, passes a light source. This causes a jet of water to blow the dead egg forward out of the hole into a container. As the disc revolves further round, the live eggs are also ejected into a separate container. The machine can not only differentiate between the dead and live eggs but counts them separately. The counting rate is approximately 40,000 eggs per hour. Other commercial machines can count and sort dead and live eggs at rates up to one million per hour. They are a useful tool for the large-scale operator and the cost, including purchase price, is less than one fifth that of manual picking.

Hatching and alevin development   Egg shells should be removed from egg baskets after the alevins have hatched out, together with eggs which have failed to hatch. The contents of the baskets may look a mess at this time and a double-bottomed egg basket which allows the alevins to drop through a removable, perforated inner-tray is an advantage. Cleaning up after hatching is more of a problem in neutral or alkaline water. Shells and other debris dissolve and disappear fairly quickly in water of a pH below 6·5.

Alevins should continue to be kept in semi-darkness until they are ready to feed. Some salmonids give a clear indication of when feeding should start by swimming up from the bottom. It is relatively difficult to judge the right moment with Atlantic salmon. Feeding should begin when the temperature is over 8°C and the yolk sacs are nearly absorbed. At this stage there is a small orange bulge remaining on the belly of the fish. They will be seen to have turned into the normal dorsal-ventral swimming attitude, and no longer lie partly on their sides due to the weight of the yolk sacs.

Supersaturation   Saturation of water with oxygen means that the water has dissolved the total volume of the gas which it is capable of taking up at a given temperature. Water, particularly from underground sources, can become supersaturated with gas due to being subjected to pressures greater than atmospheric. Water taken from below the turbines in hydro-electric stations or from the stilling basin below the spillways in high dams can also become supersaturated. The usual gases involved in supersaturation are nitrogen (air) and carbon dioxide. The effect on fish, particularly fry, is to produce what is know as gas-bubble disease when bubbles of gas appear under the skin. Carbon dioxide can also cause a condition known as nephrocalcinosis in which calcium is deposited in the fish's kidney.

The surplus gas can be removed from the supersaturated water by exposing it to air in a thin layer. The correct balance is maintained in fast-flowing rivers by the water running over the shallows. In a hatchery, the surplus gas can be removed by creating artificial falls or percolating the water through perforated metal plates. Mechanical de-aerators may have to be employed in large hatcheries using underground water. Blowers force great masses of air through shallow tanks. Such arrangements are fairly common in cold climates where bore-hole water is frequently used for hatcheries, initial fry rearing and over-wintering of parr.

Hatchery design (see   The standard Danish-type trough and egg box or
Fig 17a,b)   basket incubation system works well for salmon. The

86

trough is 215cm long, 40cm wide and 17cm deep
(7ft × 15$\frac{1}{2}$ × 6$\frac{1}{2}$in approx). Each trough holds four square
boxes which fit across the width. The bases and upper
half of the downstream ends of the egg boxes are
perforated. Water flows under the upstream end of
each box, up through perforations in the base, through
the eggs, out through the perforations in the
downstream end and so on from one egg box to the
next.

Under average conditions, each box holds about
7,000 eggs and a standard trough will incubate and
hatch 25–30,000 salmon eggs. If the eggs are intended
for sale and are only incubated to the eyed stage, the
capacity can safely be doubled but a fungicide may
have to be used. The water supply should be perfectly
clean, saturated with oxygen and have a pH between 6
and 7·5. The troughs must be covered or the hatchery
building kept in darkness.

Fig 17 Salmon hatchery—
200,000 egg capacity

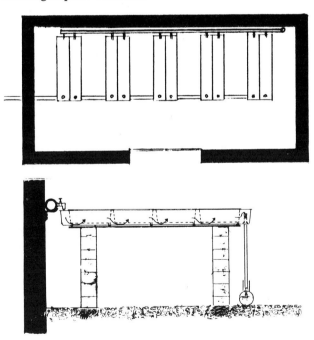

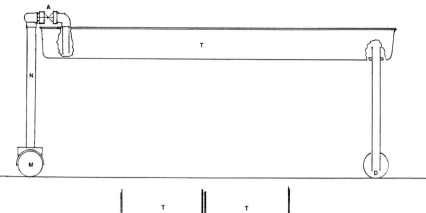

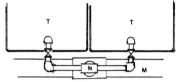

Fig 17b Water supply and
drainage – hatchery
troughs. T = troughs;
M = intake main;
N = intake to troughs;
A = valve; D = main drain.
Approximate dimensions:
length 2130mm, width
406mm, depth 179mm
(7ft × 16in × 7in)

The model hatchery illustrated is designed to
incubate and hatch 200,000 salmon eggs. The surface
water supply is delivered at an assumed hydraulic head
of 1 in 100 through a 100mm (4in) UPVC pipe. Each
trough has a separate water supply taken from the
100mm main through a 25mm (1in) UPVC pipe and
ball valve and is drained by a 50mm (2in) vertical
UPVC pipe passing through a flange at the downstream
end. The water level in the trough is adjusted by
raising or lowering the vertical pipe. The pipes from
the troughs discharge into a 150mm (6in) PVC drain
pipe. The interior dimensions of the hatchery building
are 9m (30ft) by 3·5m (12ft). The building is floored
with 20cm (8in) of river gravel on a hardcore base.
There is a wide sliding door on the long side of the
building which is insulated with polystyrene sheeting.
Electricity is laid on and a small industrial fan heater is
mounted on one wall.

6 Smolt culture

The international development of salmon farming has brought about specialization. Many sea farmers who grow salmon to market size do not have a shore site with sufficient fresh water for smolt rearing or the necessary skills. Smolt culture requires a different technology to that of growing on to market size. A higher level of ability and skill is required than for work on sea pens or cages. A sea farm site producing 100 tonnes of salmon a year needs only 50,000 smolt.

The capital outlay and running costs of operating a small smolt unit may prove uneconomic, particularly if it is far from the sea site. Salmon farming companies with several separate sea farms can afford to set up a large, central smolt farm; but salmon farmers producing less than 200 tonnes of fish a year may find it more profitable to buy in their smolt from specialist producers.

A branch of the salmon farming industry specializes in producing eggs from farm stock of known genetic make-up but smolt farmers generally prefer to maintain their own genetically stable stock. Arrangements are made with a sea farming customer to grow a cage of brood salmon on a regular basis so that the smolt farm continues to use eggs from its own fish.

The original basic system of commercial smolt culture involved the use of running fresh water, possibly from a spring or bore-hole, for incubating eggs and an intake from a stream or river for on-growing parr. Where possible, feeding fry are now transferred to cages floating in a lake as soon as they are big enough to be retained by the smallest mesh net. The capital cost of cage production is much less than for a running water unit. There is less danger of pollution or failure of the water supply, less risk of disease and the fish generally do better because they are not stressed, as they can be, in shallow tanks.

The objective in smolt culture is to introduce the largest possible smolt into the sea in the shortest possible time. Smolt farmers using heated water through a heat exchanger, are now able to produce S0 Atlantic salmon smolts. That is, fish which are ready to go to sea in less than one year after hatching. The preference for large smolt can be independent of their age. Large smolt do better when they go to sea and take a shorter time to grow to market size than smaller smolt that may be a year younger. It is more profitable in terms of fish feed and labour costs to grow the fish on for an extra year in fresh water.

Fry production The same initial procedures apply to fish which are reared on to the smolt stage in tanks with a running water supply or are transferred as large fry or fingerling parr to static cages in a lake.

Fry tanks Some hatcheries incubate and hatch their eggs in large egg trays suspended at the surface of the water in the tanks which are used for fry rearing. A removable plastic substrate, such as artificial 'grass' can be laid over the base of the tanks to give shelter to the alevins during the yolk-sac stage and at start-feed before the fry begin to swim up. The tanks cannot be cleaned while the substrate is in position and casualties, which

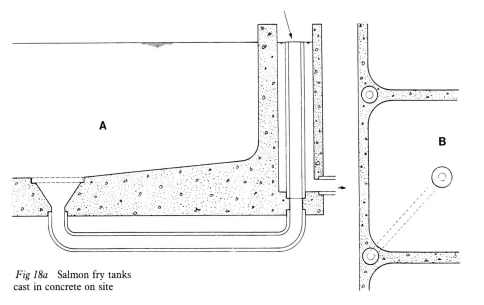

Fig 18a Salmon fry tanks
cast in concrete on site

are difficult to see and remove, can become centres for fungus infection. Many salmon hatcheries prefer to transfer the fry at swim-up directly from the hatching troughs to the bare tanks and to adjust the flow so that the fish can maintain position without being stressed.

Small tanks, either round or rectangular, may be used at start-feed and the fry transferred to larger tanks at swim-up. Hatcheries and fry units intending to move the fish into floating cages usually use standard 2×2m square tanks or round tanks with a base area of about 7m^2. Round tanks are sometimes preferred because they are believed to circulate the water more evenly but the standard square-shaped tanks are equally satisfactory provided the corners are well rounded. Square tanks take less space and can be fitted together in groups of four which simplifies the water supply system and allows each tank to be approached from two sides.

Water quality The water must be clean and free from peat fibres or

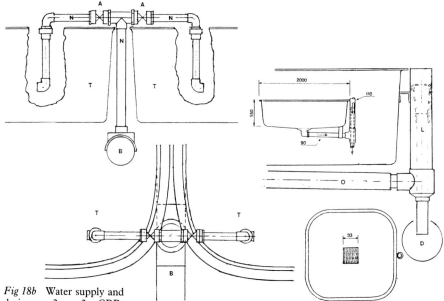

Fig 18b Water supply and
drainage – 2m × 2m GRP
tanks. T = tanks;
B = intake main; N = intake
to tanks; A = valves;
O = outlets from tanks;
L = water level control;
D = main drain. All
measurements are in mm

other solids in suspension. If not, it must be
adequately filtered. Water temperature should ideally
be in the range of 10–15°C. The oxygen content
should be at least 90% of saturation at 18°C
(8·8 mg/l).

Water flow and fish
density

The useable area of tank base is about two thirds of
the total area in a 2 × 2m square tank and rather more
in 3m diameter round tanks. The total weight of fry
that can be held per m^2 of tank area depends on the
flow which can be given without the fish being swept
away or stressed as a result of the energy they have to
expend in maintaining position.

The temperature of an unheated spring or surface
water supply is unlikely to be more than 10–12°C at
the time of year when feeding starts and the fish are
transferred to fry tanks. At this stage 10,000 fry can be
stocked per m^2 of useable tank base area. The flow
into a 2 × 2m square tank or a 3m diameter round tank

92

should not be more than about 20–30 litres/min (4–6 gallons).

By the time the fish have grown to an average weight of 1g, the water temperature can be expected to be about 12–14°C and the stocking density will have to be approximately halved at the same rate of flow. A slightly higher stocking density can be maintained if the arrangement of inlet pipes and outlet screens permits a greater flow to be given without stressing the fish.

If the fry reach an average weight of 2–3g before they are transferred for on-growing to the smolt stage in larger, circular tanks; or before being moved into floating cages, the water temperature may rise to 15–16°C and the fish density should be reduced to about 12,000 fry in each 2×2m tank or 3–4000 per m^2 of useable tank base area. The flow may have to be increased by 20–25% if the water temperature rises to 18–19°C. At all stages of growth it is essential to keep a constant check on the condition of the fish. If they are showing signs of distress the flow should be increased to the maximum the fish can comfortably tolerate. It may be necessary for economic or other reasons to go on rearing the young salmon through the late summer and autumn or fall and to over-winter them in the fry tanks. A proportionately greater total weight of fish can be kept in the tanks as their average size increases but the weight per m^2 of useable tank base should not exceed 8–10kg per m^2 in either square or round tanks.

Feeding salmon fry Starter feeding is often given in the hatchery troughs by hand. Automatic feeding is essential as soon as the fry are fully on the feed. It is important to avoid the stress of a human presence in juvenile fish, as well as to maintain regularity in the nutritional regime. Many different types of automatic feeders are available. The feeder should be capable of being adjusted to release a given amount of feed at pre-determined intervals and

93

turning itself off in darkness and on in daylight. The best feeders can distribute the food over the tank surface rather than dropping it in one place. Most models are powered by electricity but the simplest and best auto-feeder now available is operated by clockwork. It is fail-safe and only needs winding up when the food contained is being replenished.

The correct intervals and duration of feeding and the quantities and grades of feed are given in the feed manufacturer's manuals.

Outdoor tank covers and shade
Tank covers made from black, polyprop net will give shade and protection from predators as well as sheltering the fry from visually induced stress. In areas where there is strong sunlight for prolonged periods, it is advisable to fit centrally hinged covers. Covers must be easy to lift and to remove and replace. Larger, deeper tanks used to grow-on fish to the smolt stage can safely be covered by black plastic nets and do not need solid covers. Provision must be made for automatic fish feeders in tank covers.

Fig 19 Parr tanks and feeders

Indoor fry tanks The cost of housing fry tanks in a purpose-built wide span building or in an open-sided barn can be justified where wind and weather make it difficult to work outside. If feeding fry are brought on early in the year when it is still cold or there is snow on the ground out of doors, a well insulated building is essential. Shelter is also necessary if fingerling parr are to be over-wintered in tanks in cold climates where the average air temperature is below freezing point for long periods.

Fingerling parr to smolt
Large tanks with a gravity water supply

The tanks most generally used for over-wintering and on-growing young salmon to the smolt stage are round, constructed from glass reinforced plastic (GRP). Atlantic salmon fingerling parr expected to make S1 smolt, or coho that will go to sea the following spring, can if necessary, continue in the 3m (12ft) diameter tanks in which they were reared as fry. Atlantic salmon graded to make S2 smolt should be transferred to 5–6m diameter tanks which can be up to 1·5m deep. A separate water supply to each tank should be taken

Fig 20 Covered tanks for over-wintering parr

95

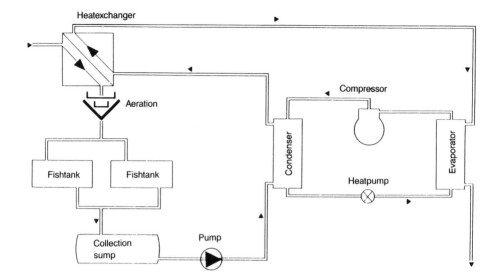

Fig 21 Heat pump for
parr tanks designed by Sea
Farm Trading of Bergen,
Norway, to reclaim heat
from effluent water. A
similar system can be used
to draw heat from the soil
and from deep fresh water
or the sea.

from a ring main controlled by a valve or T junction
from the incoming gravity pipeline.

Where possible the intake should be in a dam at the
outlet to a lake, below the depth to which ice is likely
to form. When the supply is taken from a stream, the
intake should be in deeper water or in a dam that
raises the water level sufficiently to submerge the pipe.
Intakes have to be screened and the screens kept clean.
A horizontal intake in a sump in the bed of the river,
covered by a flat screen, is partially self-cleansing.

Pre-S2 Atlantic salmon grown-on through a second
summer and winter will need a flow of about 100
litres/min (20 gallons) for each 1,000 fish. Really good
S2 smolt, over-wintered in water warm enough to keep
them feeding well into the autumn or fall and to start
again early in the spring, can make average weights of
over 120g, but the additional weight of fish per m^3 of
water will not require more water at the lower
temperatures.

The tanks will have to be covered by shade nets and

96

fitted with automatic feeders, either suspended over the tanks on a horizontal cross-bar or mounted on the side wall, but capable of ejecting the feed over a wide area. A useful design modification is to have two outlet pipes from each tank, one of which leads to the main drain and the other to a screened sump. Smolt can then be passed through the auxiliary 'fish' pipe and extracted from the sump, thus avoiding stress and possible scale loss or other damage caused by netting the fish from

Fig 22 Smolt rearing tanks

Fig 23 S_1 salmon smolt

Fig 24 Grading S₁ smolt

the tanks. The water carrying the fish from the tanks is pumped out of the sump while the smolt collect behind the screen.

Pre-S2 Atlantic salmon parr do not do as well in tanks as in floating cages unless the water supply is taken directly from a relatively warm lake where the temperature remains fairly constant, without much diurnal variation between day and night, or over-heating in sunny weather. A substantial and completely reliable flow of clean well-oxygenated water is essential. A tank unit designed to produce 100,000 S1 and S2 smolt (about 50% of each) needs a minimum, guaranteed flow of at least 4mgd (4 million gallons in 24 hrs) or $0 \cdot 2m^3$/sec.

Floating cages Cages, given the right conditions, cost less and produce bigger and better smolt than fish reared in tanks with an unheated water supply. Salmon fry are reared in

98

small fry tanks until they are big enough to be transferred to floating cages anchored in a lake. The gravity water supply is a fraction of that needed to grow the fish to S1 or S2 smolt in tanks and the capital cost of cages is a great deal less than the cost of site preparation, water intakes, pipelines, large tanks and plumbing.

Atlantic salmon fry have been successfully transferred for on-growing to the smolt stage in floating cages in freshwater lakes when their average weight has been barely 1g. Husbandry is difficult when the fish are so small and they are much easier to look after if they are not transferred until they weigh about 3g. The limiting factor is a cage-net mesh size small enough to contain the fish. Very small mesh nets have to be changed at frequent intervals. They become coated with algae in warm sunny weather and no longer allow sufficient interchange of water. The net changes stress the fish, putting them off their feed and slowing down growth. It may be better to wait until the average weight of the fry is at least 2–3g before they are moved from fry tanks to floating cages.

Lakes   Good lakes for smolt cages are generally on low ground, close to sea level, with a western or southern exposure and in areas having an equable, maritime climate. The best sites are in fairly shallow lakes in rolling country or low hills not in rocky, mountainous terrain. The water must be perfectly clean and saturated with oxygen. It should preferably be neutral or slightly acid as this discourages algal growth and is a less advantageous environment for bacteria than alkaline water. The temperature in the top 3–4m of water in the lake should warm up to 10°C early in the summer, not go above 16–18°C in hot weather and remain above 6°C late into the autumn or fall. The lake should be ice-free in winter or only form thin ice for short periods. The relatively stable, diurnal temperature of the water in a lake, compared to

99

running water taken from a stream, is of great advantage in promoting quicker growth in cage fish.

Sites   A cage site should be sheltered and not exposed to a fetch (length of open water to shore) of more than about 2km ($1\frac{1}{4}$ miles), preferably less in the direction of the prevailing wind. The depth over the site area should not be less than 6m (20ft) or greater than about 15m (50ft). The bottom should be level or gently

Fig 25   S$_2$ salmon smolt

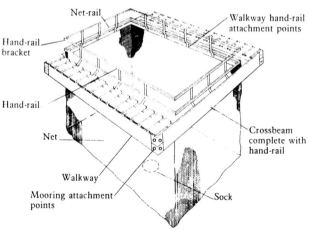

Fig 26   The Kames floating fish cage

sloping sand or gravel, not boulders or rock. The cages should be not more than about half a mile from a shore base where a jetty or floating walkway can be constructed and a boat or service raft safely moored. The shore base should be easily accessible from the nearest road.

Smolt cage design   Small cages holding 1–3 tonnes of fish and enclosing 50–130m^3 of water are better than large cages for on-growing young salmon to the smolt stage in lakes. The net cages should be supported by a rigid, rectangular flotation collar with wide walkways providing stable working platforms. The best type is the well-known Kames cage, manufactured in Scotland, which has 6×6m collar supporting a net cage $4 \cdot 4 \times 5 \cdot 8 \times 5$m deep ($20 \times 20$ft collar, $15 \times 18 \times 17$ft deep net).

Cage nets are made up in the form of a bag from knotless nylon netting. The net can be made to taper down to a codend, closed with a cord and pin, that can be opened to take out accumulated debris and any dead fish. This allows an accurate count to be kept of casualties. The cage nets are easy to raise by nylon ropes passing round under them sewn to the mesh.

Top nets and guard nets are fitted over and around the rails on the walkways, above the water, to keep out predators. Nets with a mesh of 3–4mm, measured from knot to knot along one side have to be used for 1–3g fry and will have to be replaced with larger mesh as the fish grow. These very small mesh nets may have to be changed at frequent intervals. If the fish are transferred to cages at an average weight of over 5g they can be safely retained in a mesh of 6mm. In the second summer when pre-S2 parr reach an average weight of 25g they can be transferred to cages with 13mm mesh.

Moorings (see Fig 27)   Conventional anchors are preferable to home-made anchors. Heavy concrete blocks can be used instead of anchors on mud or clay bottoms.

A For sheltered sites and up to six cages, a single 150kg (330lb) anchor on depth ×3 of chain with a swivel at each end should be adequate. Shackle the surface end of the chain to a 75kg (approx. 165lb) buoy until the cages are connected. The surface chain is 13mm ($\frac{1}{2}$in) with 50mm (2in) links, shackled to the top of the mooring chain. Support the surface mooring chain on 36kg (80lbs) buoys at 6m intervals until the cages are pinned to the surface chain. The large buoy can then be removed. This is a swinging mooring which allows the cage flotilla to move into the wind and prevents faeces and waste food collecting in one place.

Fig 27 Mooring layouts for smolt cages in lakes

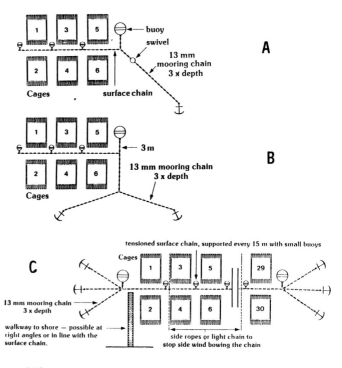

102

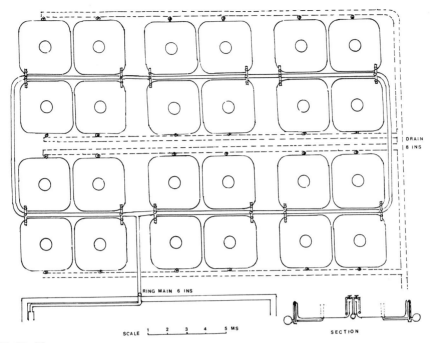

DRAIN
8 INS

RING MAIN 6 INS

SCALE 1 2 3 4 5 MS

SECTION

Fig 28 First summer rearing: this unit will grow 150,000–200,000 parr to an average weight of 3–5g depending on water temperature

B   For less sheltered sites with a longer fetch and six to eight cages. It is a swinging mooring similar to A, rigged and mounted in the same way, but with two anchors attached to a single bridle.

C   For sites where ten or more cages can be moored. Access to the flotilla can be by pontoon from the shore or by boat. Rigging and mounting is similar to A and B but three anchors on separate chains are linked to large buoys at each end of the surface chain.

Pacific salmon fry and smolt

The detailed procedures for rearing Atlantic salmon smolt are generally applicable to the species of Pacific salmon that must spend an appreciable time growing in fresh water before they are able to move into the sea. The length of time taken and the rate of growth varies according to the particular species.

103

The freshwater growth of coho, the species which has so far proved to give the best return to salmon farmers, can be very fast. In temperate climates, where the temperature of an unheated water supply warms up to 11–12°C by May, young coho can reach a weight of 18–20g during the first summer. Some casualties have been experienced with smaller fish put out to sea cages in August but later transfers with larger fish of 30g average weight have resulted in insignificant losses. The smaller fish can be acclimated to sea water if they are first introduced to water of 10‰ salinity for a period of four to six days. The salinity is then increased to 20–25‰ over the next 10–20 days according to whether the fish show any signs of distress. They can then be safely transferred to sea cages in water of full marine salinity (30–35‰).

Atlantic salmon smolt—outline production schedule

Year 1

Hatchery	650,000 'green' (newly fertilized) eggs less losses 540,000 start-feeders
Fry tanks 36. 2 × 2m	540,000 feeding fry less 15% losses 459,000 feeding fry

June
Grade 1g or over transfer to 4mm mesh nets in 3 tonne cages 36,000 per cage approx. (7 cages)	206,550 (45%) probable pre-S1
Under 1g retain in tanks	252,450 (55%)

July
Split cage stock at 3g	206,550

10,000 per cage approx. (25 cages)	probable pre-S1
Split tank stock 1g or over to 4mm mesh cage nets (4 cages)	114,700 (45%) 137,750
Under 1g retain in tanks	probable S2 and runts

October

Grade cage stock over 10–12g to 6mm mesh nets 7000 per cage (46 cages)	321,300 pre-S1
Grade tank stock over 1g to 4mm mesh nets 15,000 per cage (8 cages)	123,980 pre-S2
Dispose of runts under 1g or overwinter in tanks	

No allowance has been made for casualties which could reduce the totals of fish at the October grading by 5–15%

Year 2

March
All smolting S1 counted and sorted for sale

May
S1 smolt graded and sold
Grade out for sale any fish in the S2 cages that may have smolted as S1
Grade S2 cages and transfer all fish to 6mm mesh nets
Ungraded new entry in $2 \times 2m$ tanks

105

7 Freshwater operations

Successful salmon farming in fresh water and the sea depends on good husbandry by skilled fishmasters. Like any other form of livestock production the well-being of the fish depends on constant supervision backed up by the means to maintain healthy and rapid growth. Expertise in day-to-day management needs to be acquired by practice and not by theory.

Grading The after-effects of stress during grading small fish can last for about 30 days and reduces the growth rate during that period. Grading of all young, anadromous salmonids, intended for transfer to the sea, should be kept to a minimum and if possible to one, or at the most two, gradings before smoltification.

Atlantic salmon need to be graded only once during the first summer when the fish are transferred for on-growing or over-wintering in larger tanks or cages in fresh water. Most of the parr which reach a length of 10cm by November of the first year will smoltify the following spring. Farms specializing in smolt production may need to measure as well as grade their pre-smolt parr to check growth rates. Measurement can be carried out electronically but the fish may have to be anaesthetized before passing through the counter.

106

All grading which involves catching the fish causes stress but it is possible to construct a simple 'crush' grader for use in round tanks which involves the minimum of fish handling. This consists of a removable screen which fits across the diameter of the tank and is hinged at the centre. The screen should have rubber flaps along the bottom and at the ends, which make continuous contact with the sides and bottom of the tank. The screen is put in position straight across the tank and one half is then slowly turned to bring the two halves close together. The larger fish which have not been able to pass through the screen are then left in a small wedge of water, from which they can be netted or pumped directly into another tank. The 'crush' method will only work in tanks with flat-screen drains without stand-pipes.

Aeration   A means of aeration should be provided to maintain the dissolved oxygen content at a safe level in fry and smolt tanks if there is a risk of partial failure of the water supply or water shortage in hot weather.

The simplest method for achieving aeration is by pumping air into the water through porous distributors. A system that can be used in hatcheries is to suck air into the water supply to each trough through a venturi on a long vertical pipe from a high-level main.

Where aeration is essential, either continuously or at regular intervals, oxygen can be introduced directly into the water from a bank of cylinders or a liquid oxygen storage tank. An alternative, which works very well in practice, is to use an oxygen generator of the Xorbox type made by Ewos Aquaculture. This takes compressed air from cylinders and separates the oxygen from the other gases. Oxygen produced in this way costs less than that bought in cylinders.

Heated water   The results of hatching and rearing young salmon in artificially warmed water can be dramatic. It is possible

107

to reduce the normal incubation period by as much as two to three months and subsequently produce fry approximately three times heavier than those hatched and reared in northern waters at ambient temperature. The cost can prove prohibitive where energy is expensive but in Iceland where geothermal resources are available there is a practical potential for producing zero-age, S0 smolts, fish which are large enough to go to sea in less than one year after hatching.

Speeding up growth in the first summer so that the fish quickly reach the size for smoltification depends on early spawning and accelerated incubation. Eggs from salmon which are ripe to spawn in late October or early in November can give feeding fry by late December or early January. The optimum temperatures are approximately 9–10°C during incubation and the yolk-sac stage then 10–12°C when the fry have started to feed. The supply of warm water must be maintained until the natural water temperature has reached about 12°C. No advantage can be gained from simply reducing the incubation time. The fish must be in water of temperature sufficiently high to induce feeding as soon as their yolk sacs have been absorbed. They should then be maintained in an environment where the water temperature is steadily increasing towards the optimum for growth in fresh water.

The ambient temperature of the incoming water supply should be continuously measured by a sensor and the heat input required to raise the temperature to the pre-determined level controlled and adjusted electronically. An alternative and safer method is to use a heat exchanger to heat the water in a separate circuit and to control the temperature by adding cold water in a mixing tank. In either method the warmed water should not be piped direct to hatching troughs or fry tanks but to a header tank fitted with sensors and an alarm system.

Countries such as Iceland which have hot springs can

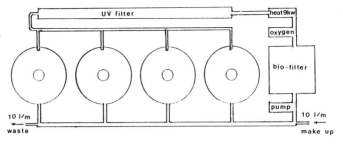

Fig 29 Diagram of re-circulation system: 210 litres/min in circulation; 10 litres/min make-up with fresh water

use the geothermally heated water to warm the water supply to salmon hatcheries and smolt rearing tanks through heat exchangers. In the Icelandic system eggs are stripped from early spawning brood stock and incubated at 10°C. The fry are fed in water at 10°C rising gradually to 13°C. The temperature is then maintained at 13–14°C. No further acceleration to the growth rate is obtained by keeping the local Icelandic race of Atlantic salmon in warmer water. Atlantic salmon smolts reared by this method reach an average weight of 35g by September of the first year.

Coho salmon respond particularly well when incubated and hatched in warmed water. Fish which start to feed by the end of January can reach a weight of 20g and be acclimated for transfer to the sea by May/June.

Heat pumps and solar heating Some hatcheries in particularly cold areas have installed heating coils in the deep sea water of nearby fjords. A heat pump then warms the water supply to the incubating eggs. The system works well but the cost of pumping is not much less than for direct heating.

Solar heating is not a practical proposition for hatcheries because there is little enough daylight, let alone sunshine, during the winter in the high latitudes, either north or south. The flow needed when the fry are feeding during the late spring and summer is too great to be warmed up by panels. Either method might

109

prove viable as a comparatively low cost source of energy where the water for a hatchery and fry rearing unit is re-circulated.

Re-circulation The economics of re-circulating the water used in a hatchery or smolt rearing unit is only practical in special circumstances. It has proved profitable in the far north where there are good saltwater sites for sea cages for on-growing salmon, but where the fresh water sources on land are too cold to produce a satisfactory proportion of S1 smolt (fish which smoltify the year after hatching).

The basic re-circulation system consists of a group of tanks to hold the fish. The flow leaving the tanks is collected in a single pipe and pumped through a biological filter. Oxygen is injected into the filtered water which is then heated. The warmed water passes through UV tubes or filters which destroy any bacteria, and is returned to the fish tanks, usually through a perforated spray bar. Fresh water must be added to about 10% of the flow in circulation and a corresponding flow drained off after leaving the tanks.

Re-circulation systems are safer if they are made up of several small, separate units. A unit of four circular tanks each with a capacity of $3 \cdot 5m^3$ of water will require a re-cycled flow of 200 litres/min with 20 litres/min added as 'make-up' water. A valve in the main collecting pipe draining the four tanks is set to drain out a corresponding flow and the fresh water is introduced at the input to the pump. Each unit will need a 9kw heater. A unit of four tanks will rear approximately 7,000 young salmon to an average weight of 40g.

Mixing salt and fresh water Atlantic salmon fry will tolerate a salinity of about 10‰ in the water supply to tanks, or in cages floating in brackish water, from a weight of 3–5g. The salinity is gradually increased as the fish grow and by the time they reach the smolt stage they are acclimated to sea water of full marine salinity. Apart from the advantage of acclimation, the fish will grow more quickly if the sea water used in the mixture is warmer than the source of fresh water during the winter months. The tanks on smolt farms in cold climate countries can be equipped with a dual fresh and saltwater supply.

Disease risk The main source of infection by diseases, such as furunculosis in Atlantic salmon and BKD in Pacific salmon, which can be carried by the fish from fresh water to the sea-pens or cages, is from wild salmon. The farm fish are at risk from the wild fish not vice-versa.

If at all possible the water supply to tanks should be taken from a source inaccessible to wild, anadromous salmonids, and they should not be present in lakes used for on-growing young salmon to the smolt stage in floating cages. A sample of the sedentary, wild fish populations in lakes, or sources of running fresh water, should be caught, tested for bacterial or virus diseases and parasites likely to infect salmon fry or parr held at farm densities. Equipment is available, if required, for pressure filtration and the UV sterilization of the water supply.

Smolting Atlantic Pre-smolt parr naturally try to store vitamins in
salmon preparation for smoltification. Fish food manufacturers make some allowance for the special needs of pre-smolt parr. These are mainly for added vitamins and fats. The A and B1 vitamin requirement is approximately doubled and for B2 can be as much as five times greater during smoltification than in pre-smolt parr. Vitamin deficiencies must be replaced in the diet. Many of the distress symptoms in smolt on transfer to salt water may in fact be attributed to a reduction in their resistance to stress due to avitaminosis.

The triggers which set off smoltification are increasing daylight and rising water temperature. Parr may start to become silvery at weights of 25–30g when

111

the temperature reaches 7–10°C and smoltify completely when the water reaches 10–12°C. The average weight at full S1 smoltification is approximately 35–40g, if the fish have been over-wintered in unheated, surface water in a temperate climate.

All S1 (or S2) smolt do not smoltify at the same time. Some fish may take several weeks longer and will not smoltify until well into the summer. The smaller parr may not begin to smoltify until the water reaches 11–12°C.

The changes in the appearance and behaviour of pre-smolt parr at smoltification are the essential indicators showing that the young fish are ready to be transferred to the sea. The most distinctive alteration in appearance is the silvering of the fish. This masks the finger-print parr markings and spots on the sides of the fish. In the early stages of smoltification the parr markings are still faintly visible but the fish become lighter in overall colour and finally the finger marks disappear completely.

At this stage the scales become very loose on smolt and are easily displaced. This is one of the factors which makes smolt particularly prone to injury and difficult to move without causing casualties.

There is one observable change in the behaviour of young Atlantic salmon, indicating that they will tolerate being transferred to salt water, which is particularly useful if there are no facilities for acclimation. The fish in round tanks, which have been swimming against the current, facing 'upstream', turn round and intermittently swim 'downstream' with the current round the tank.

Road transport of smolt Fully developed smolt are delicate and difficult to transport without physical damage and stress. Tanks, even for comparatively short distance transport (100–200 miles), should be insulated. The fish are safe to travel in water at the ambient temperature of the

112

supply to the rearing unit in cool weather. The water temperatures in the transportation tanks should be equal to that in the tanks at the delivery end before the fish are transferred. The tanks on the delivery truck should have a supply of pure oxygen (spare cylinders should always be carried), and of air from a petrol-driven compressor. A water pump is also essential on longer journeys where the water may have to be changed.

The fish tanks on trucks used for long distance transport should not only be insulated but also have a thermostatically controlled cooling system. The fish are put into the water in the transporter tanks at the same temperature as that in their rearing tanks. The water temperature is then slowly reduced to 4–5°C and kept at that temperature during the journey to render the fish less active and reduce their oxygen demand. A re-circulation system incorporating a filter may also be necessary for long deliveries. Some losses through stress are inevitable even when every precaution has been taken.

Acclimation Salmon, even when fully smoltified, are more safely transferred to sea water if they are given a period of acclimation in gradually increasing salinity. If the smolt rearing unit is close enough to the sea, a saltwater supply can be laid-on directly to the large parr tanks. The seawater intake should be similar to that described for shore-based on-growing tanks.

Pink salmon fry, which naturally migrate to salt water as soon as their yolk sacs are absorbed, need a seawater supply laid-on to the initial feeding tanks. They have to be grown-on in shore-based tanks until they are large enough to be transferred to sea cages, depending on the minimum practical mesh size that can be used. The saltwater supply to the shore unit has to be in the ratio of about 20 to 1 to the freshwater supply.

The arrangement of special acclimation tanks will

113

depend on the site but the water supply should be by gravity from header tanks at the highest level which can be kept full by pumping. Mixing can be done in an intermediate reservoir, but a better system is to have separate fresh and saltwater mains providing a gravity flow to large diameter GRP tanks, each with a separate intake and mixer valve.

Some of the most successful salmon farms have dual fresh and saltwater supplies to their smolt rearing units. The parr are gradually acclimated to an increasing degree of salinity over a period of several months prior to smoltification. They can then be transferred directly to sea water of 30–35‰ without stress or loss.

Smolt delivery by well-boat

Well-boats of several hundred tonnes designed for transporting live marine fish and eels have been adapted to deliver smolt long distances by sea. They are useful for taking fish out to off-shore islands or to and from sea cage anchorages that have no proper access by road. The fish are loaded at relatively low density in a hold filled with well-oxygenated sea water. Smolt have to be counted when they are taken on board and again as they are being delivered at the sea farm. Live fish transport by well-boat is expensive and the cost is usually borne by the smolt producer although it is passed on to the sea-salmon farmer.

Floating fish tanks

If a marine site is not accessible to delivery by well-boat, moving smolt out to sea cages can present problems when they are a long way from the shore base. It is seldom possible to bring even a small flotation collar with a shallow net close enough inshore for direct loading from a live-fish transporter. Ingenious salmon farmers have solved the problem by constructing their own miniature well-boats. These generally consist of a flat-bottomed, hollow hull made from polystyrene filled GRP, about 7m in length and 4m wide. Sea water circulates through the central

114

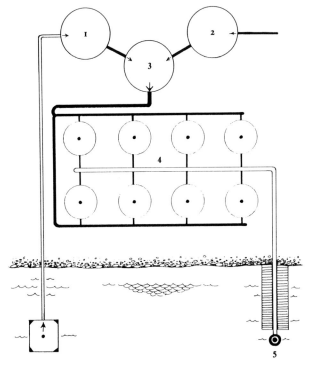

Fig 30 Smolt acclimatization: 1 seawater header tank; 2 freshwater header tank; 3 mixing tank; 4 smolt tanks; 5 smolt collecting point for transportation to cages

open-well through fish screens in the bow and stern.

Smolt age The symbols S0, S1 and S2 used by Atlantic salmon farmers simply refer to the age of the fish when they are ready to go to sea. S0 means fish which have smolted in less than a year from the date they hatched and so on. The production of S1 or if possible S0 smolt is regarded as highly desirable. It is worthwhile considering what happens to the fish after they go into sea cages. The bigger the smolt are when they are moved into salt water the sooner they will grow-on to market size and the quicker will be the turnover in the marine unit. S2 smolt weighing 100–150g may grow to a marketable weight of 3kg after less than one year in the sea. Ordinary S1 smolt weighing about 40g may only reach a weight of 1kg before maturing as grilse and having to be slaughtered.

115

8 Atlantic salmon sea farm project

The projected development is for a salmon production
or 'fattening' unit for on-growing salmon to market
size in floating cages in the sea. The young fish are to
be bought in at the smolt stage, fully acclimated to salt
water. The capital and running costs of a marine
salmon farm are governed by the sea site, its depth,
tidal flows and shelter, and the accessibility of the
associated shore base. The cage site for the projected
sea farm meets all the essential criteria. The shore base
is 100m from an existing public road and the
intervening ground will not require the construction of
drainage culverts or rock blasting to make a road to the
shore.

OUTLINE OF INVESTMENT
Only the principal items are listed.

Shore base Road construction, site preparation and jetty or
slipway.
Dry store, tool store and generator house.
Fish food silo or store, fish house, boxing and icing
equipment.
Generator and ice-maker and fuel tanks.

116

Submersible pumps, cable and hose.
4 wheel drive 'pick-up' vehicle.

Marine site   *Year 1.*
6 rectangular collars, 4 working platforms and fixed moorings.
12 rectangular cage-nets, fence-nets, cover-nets and 2 predator-nets (all round the cage flotilla).
6 automatic feeder units with control gear and batteries. Battery charger and spare batteries. Food storage bins.
Work-boat.

Year 2. Contingency only.

Year 3.
4 octagonal plastic collars, 4 working platforms and fixed moorings.
4 octagonal cage-nets, fence-nets and cover-nets.
Additional predator-nets.
High-pressure water-pump and 4 water-jet feeders.

Year 4.
2 octagonal plastic collars, 5 working platforms and fixed moorings.
4 octagonal cage-nets, fence-nets and cover-nets.
Additional predator-nets.
Replacement nets contingency.

Fixed costs   Most items will depend upon local circumstances but the figures for depreciation and essential labour have general application.

Depreciation.

Flotation collars. Metal or wood	8–12%
Plastic pipe	6%
Working platforms	10%
Cage-nets	25%
Other nets	20%
Moorings	20%
Shore-based buildings	5%

117

Generators and dry machinery	10%
Pumps, pipes and valves (corrosion resistant)	10%
Boats	10%
Vehicles	16%

Labour and management
It is assumed that on-site supervisory and financial management will not be provided and that the farm will be in the charge of a trained and experienced working salmon farm manager.
Year 1. Working manager, 1 salmon husbandryman, 1 semi-skilled man.
Year 2. Working manager, 1 salmon husbandryman, 2 semi-skilled men.
Year 3. Working manager, 2 salmon husbandrymen, 2 semi-skilled men.
  An assistant salmon farm manager or senior husbandryman may be required when the salmon farm is in full production.

Variable costs
All these items depend on local circumstances. Stock insurance is generally dependent on the location of the shore base and the exposure of the marine site.

Profit and loss
The profitability of a salmon farm in any climatically suitable part of the world depends mainly on the cost of fish food and labour at the location concerned.

FIVE YEAR PROJECTION—stocking and production. (See salmon sea cage plans page 120)
The projected fish losses are 10% spread over the first year and the grilsing rate (fish which become sexually mature after 1 year or less in the sea) is 50%.

Financial year 1.
Stocking plan. April/May
October

Fig 31a
20,000 smolt in cage A (10% loss)
19,000 post-smolt in cages A, B and C

Financial year 2.
Stocking plan. April/May

Fig 31a, b
30,000 smolt in A
18,000 grilse in B, C, D, E and F

Fish marketed. June to October	9,000 grilse at 2kg = 18 tonnes
Stocking plan. October	29,000 post-smolt in A, B, C and D
	9,000 pre-salmon in E, F, G and H
Fish marketed. January to March	4,500 salmon at $3 \cdot 5$kg = $15 \cdot 75$ tonnes

Financial year 3.

Stocking plan. April/May	Fig 31b, c
	50,000 smolt in A
	27,000 grilse and pre-salmon in B, C, D, E and F
	4,500 salmon in G and H
Fish marketed. June to October	13,500 grilse at 2kg = 27 tonnes
	4,500 salmon at $4 \cdot 5$kg = $20 \cdot 325$ tonnes
Stocking plan. October	47,000 post-smolt in A, B, C and D
	13,500 pre-salmon in E, F, G, H, J and K
Fish marketed. January to March	6,750 salmon at $3 \cdot 5$kg = $23 \cdot 6$ tonnes

Financial year 4.

Stocking plan. April/May	Fig 31c, d
	50,000 smolt in A
	45,000 grilse and pre-salmon in B, C, D, E, F and G
	6,750 salmon in H, J and K
Fish marketed. June to October	22,500 grilse at 2kg = 45 tonnes
	6,750 salmon at $4 \cdot 5$kg = $30 \cdot 4$ tonnes
Stocking plan. October	47,000 post smolt in A, B, C and D
	22,500 pre-salmon in E, F, G, H, J, K, L and M
Fish marketed. January to March	11,250 salmon at $3 \cdot 5$kg = $39 \cdot 4$ tonnes

Financial year 5.

Stocking plan. April/May	50,000 smolt in A
	45,000 grilse and pre-salmon in B, C, D, E, F and G
	11,250 salmon in H, J, K, L and M
Fish marketed. June to October	22,500 grilse at 2kg = 45 tonnes
	11,250 salmon at 4.5kg = $50 \cdot 6$ tonnes
Stocking plan. October	47,000 post smolt in A, B, C, and D
	22,500 pre-salmon in E, F, G, H, J, K, L and M
Fish marketed. January to March	11,250 salmon at $3 \cdot 5$kg = $39 \cdot 4$ tonnes.

a

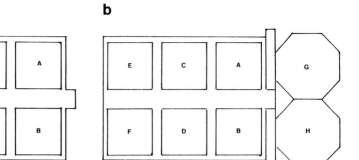

b

c

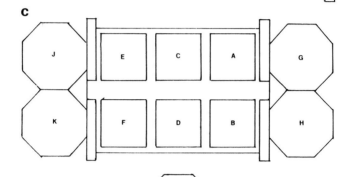

d

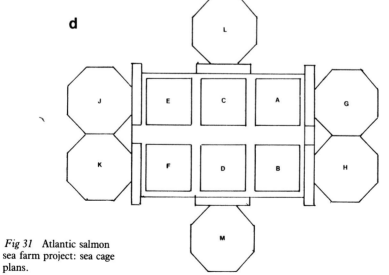

Fig 31 Atlantic salmon
sea farm project: sea cage
plans.

120

It has been assumed throughout the 5 year projection that there is a 50% 'grilsing' rate in the sea farm stock, *ie* that 50% of the on-growing stock become sexually mature and therefore have to be slaughtered and marketed as grilse during their second year in the sea. Grilsing rates as low as 5–10% have been achieved using genetically stable broodstock, either bred on the farm or from stocks derived from particular rivers. It is not unusual to achieve grilsing rates of 20–30% but a relatively high grilsing rate has been deliberately used in this projection.

9 Saltwater salmon farms

The salmon farming industry on the Pacific seaboard of
North America and on both sides of the North Atlantic
has followed much the same course. Atlantic salmon
culture is developing in the coastal waters native to the
Pacific species and Pacific salmon are being farmed in
Europe. Much the same techniques are being used
wherever salmon are being cultured. Some are more
profitable or practicable in particular locations, or may
have to be adapted to the needs of the species being
cultured, but they come within one or other of the
basic systems. These comprise tanks or ponds on shore
with a salt water supply pumped from the sea;
enclosures in the sea in bays or fjords or in channels
between off-shore islands; and floating cages or pens
anchored in more-or-less sheltered water fairly close to
the shore. Cages are so far proving to be the most
satisfactory and profitable way to fatten salmon for the
table market.

Sea farms on shore The attraction of building a saltwater salmon farm on
the shore, rather than using cages or enclosures in the
sea, is that it can be easily serviced and is not open to
damage or put out of reach by wind or weather. The
principal disadvantages are the high capital cost of

122

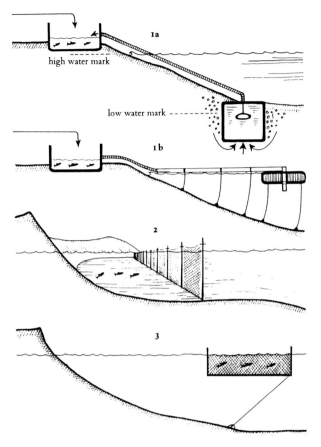

Fig 32 Basic systems of salmon farming in sea water; 1 pumping to shore tanks (a) from beach well (b) from floating intake; 2 shore enclosure; 3 floating cage

high water mark

low water mark

1a

1b

2

3

construction, the problems of corrosion and marine fouling in pumps and pipelines, the risk of pump failure, and above all the cost of continuously pumping a very large flow of water. The height to which the salt water has to be pumped is the most important factor in deciding whether a project of this kind is likely to prove worth while on a particular site. The difference between high and low tide on different sea coasts can range from less than a metre (3ft) in Norway to 18m (60ft) in such places as the Bay of Fundy in the Canadian Maritimes.

123

The danger of inshore farming salmon is from self-pollution. Tidal movements along the shore create a serious risk of pumping back some part of the effluent from shore tanks, however far the intake is from the outfall. The effluent may be almost totally diluted and the water re-oxygenated, but it can still return bacteria causing diseases such as furunculosis and the whole gamut of virus organisms capable of causing epizootic outbreaks in salmon held at farm densities in shore tanks. The volume flow required makes it economically impracticable to use either UV or microfiltration.

Saltwater supply The seawater intake is the most important part of the unit. This must be kept as simple as possible. If the sea area is well-sheltered, a floating intake is possible. This consists of an anchored raft connected to the shore by a flexible, armoured hose supported on floats. The seaward end of the hose passes down through the raft and is screened by a large rose or a tubular screen attached to the underside of the raft. Screens or roses must be removable for cleaning and be replaceable by duplicates. Pumps are electric, either submersible on the end of the hose below the raft or sited on the shore. In either case, a diesel stand-by generator with automatic switching is essential.

An alternative system is to construct a fixed sump on the sea bed, supporting a platform above the surface. The water intake can either be through side-screens above the bottom or downwards through a flat screen on the top of the sump. A number of separate smaller intakes is a much safer arrangement than a single large intake. A stand-by intake should be provided to allow for shutdown during maintenance. The piping is laid on the sea bed or preferably buried in a trench. This type of intake is more expensive than the floating type but is essential in an exposed sea-site.

Beach pumping This is a low-cost method of saltwater salmon farming which has so far only proved workable on a small

124

scale. The system is based on pumping the main seawater supply from a shallow well on the beach, between the high and low-water marks. The beach acts as a filter sufficiently fine to exclude the sporophyte and gametophyte generations of marine algae and the free-swimming, veliger larvae of mussels. The well is lined and is about 1m in diameter and 2m deep. It fills from the bottom upwards with water drawn in from the surrounding sand and gravel. The surface of the beach round the sealed well-head is kept clean by the ebb and flow of the tide and if properly constructed the well and filter area does not need back-washing.

Saltwater tanks   The basic designs are much the same as those used for salmonids in fresh water. The cheapest and simplest systems in Scandinavia consist of large ponds which usually have concrete sides. The dimensions are approximately 30m long by 10m wide and 2m deep. The bottom should slope from the long sides to the centre and towards the seaward end. The ponds should be capable of being drained dry. More than one intake should be provided to prevent the formation of 'dead' areas with no interchange of water. This type of tank will hold up to 20kg of fish per cubic metre of water and requires a flow of about 300 litres per minute.

Other methods include circular ponds and tanks in concrete or GRP, round-ended circulating raceways of the Foster-Lucas type or rectangular circulating raceways. The decision as to the type to use depends on the following factors:

— site area, the tidal range and the seawater intake;
— required flow of water and pumping costs;
— weight of fish per cubic metre of water which can be grown-on and the rate at which they will grow;
— husbandry—is it easy to distribute fish food, keep the tanks clean and catch the fish for grading?
— maintaining the health of the fish.

In selecting or designing any circulating tank or

125

raceway it should be borne in mind that the fish will burn off energy and not put on weight so quickly in a fast flow.

The materials used for making shore tanks should have the greatest possible resistance to the corrosive effects of sea water. They should also provide a surface on which it is difficult for marine plants and animals to find attachment. Concrete, reinforced with either steel or glass-fibre is not a particularly good material. Even when very well finished it still offers a relatively rough surface. A better material for tanks is either GRP or a tough plastic which is strong enough to be used without reinforcement.

Enclosures   An arm of the sea, or channel between the shore and an island, is closed off by solid, fish-proof barriers. These are usually made of steel-bar screens supported between reinforced concrete piers. The barriers must extend at least 1·5m above the surface at high water on spring tides and it is unlikely to be economically worth while to plan an enclosure of this kind in a place where there is a maximum tidal range of more than 1–2m.

The tidal flow through a fixed enclosure may be insufficient to maintain the oxygen content of the water or to remove waste matter. In these circumstances, it is necessary to use impeller pumps to supplement the tide and draw water through the enclosure, if it is to hold a high enough density of fish.

This system has been successfully developed for Atlantic salmon culture in Norway. Two enclosures are needed to cover a sea-feeding cycle of about 18 months from smolt to market size. The density of salmon (av. wt. 1·5–6kg) held in the Norwegian permanent enclosures (1·2 hectares in area) is about $0·5kg/m^3$ without pumping, and $2–3kg/m^3$ with pumping. The fish are fed by piping a wet food mixture into the water and are recaptured for market by putting a net round the feeding area. A problem is that once smolt are released into a large enclosure of the permanent

126

Fig 33 Atlantic salmon sea farm in Scotland

type the fish cannot be properly checked or treated for disease. A major difficulty in this method of salmon culture is the accumulation of waste food and faeces which may eventually fill the enclosure unless it is swept clean by the tide.

Sea cages   Salmon sea farming in both Europe and North America has, in recent years, turned increasingly towards the use of floating cages anchored in sheltered water. The majority of the new generation of sea farms are of this type. The principal advantage of cages over any kind of static enclosure, either on land or in the sea, is their versatility. They can be of the shape and size best suited to the location and the scale of the project or moved to a new marine site without incurring large capital losses.

    Salmon cages basically consist of a simple net bag suspended from the surface. A variety of materials have been employed to construct the flotation collars supporting the cage nets. These include plastic floats or blocks of polystyrene foam on wood, GRP or metal frames. Some are made from large diameter plastic pipes filled with plastic foam, or inflated hose. Cages

127

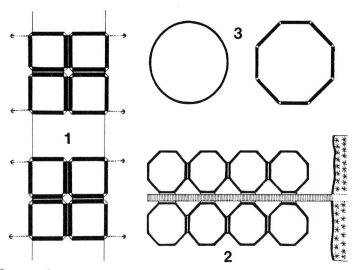

Fig 34 Sea cages: 1 flotillas; 2 cage dock; 3 round and polygonal cages for individual anchoring

can be anchored separately to the sea-bed or attached together in groups to form flotillas. They can be arranged round an anchored raft or working 'island', or moored alongside pontoons forming a walkway from the shore where there is deep water close to land and not much tidal rise and fall.

Many types of cages are commercially available. The best designs have been tested under working conditions on salmon farms by the manufacturers. They are capable of functioning well, provided they are operated under conditions similar to those in which they were developed, but they can prove very unsatisfactory in different conditions or on sites more exposed to wind and wave action. Manufactured cages fall into three main groups.

Inflexible cages The most common type is rectangular. It consists of four walkways joined together to form a flotation collar from which the cage net is suspended. The sections of walkways are pressure-cured wood or metal frames, enclosing polystyrene flotation blocks, fixed rigidly together at the corners. Vertical stanchions, joined by

128

horizontal bars or rods, are bolted to the inside of the walkways to act as handrails and to support the cage net.

One model of a large rectangular cage is made up of light metal walkways in frames supported above the water on large, separate floats. Some of the walkways are sufficiently bouyant, wide and stable to allow a vehicle to be driven between the cages. A group of rectangular flotation collars and cages can be linked together to form a raft or flotilla incorporating an integral working platform for pumps, compressors and food store.

Large flotation collars   Circular and hexagonal collars made of heavy gauge, high-density plastic tubing have been developed in Norway. The tubes are approximately 200–250mm in diameter, filled with plastic foam. Upright tubes supporting a handrail are welded at intervals to the flotation collar. The main collar is formed by two concentric circles or polygons of tubing, joined at intervals by welded distance pieces. They support a continuous walkway round the collar. The cage net is attached to the inner flotation tube and a 'jumper' net to stop fish leaping out of the cage extends up to the handrail.

A variation of the circular collar has a wide central walkway with handrails across the diameter of the circle. This allows two net cages to be used, one on each side of the walkway.

Cage nets supported by circular or polygonal plastic collars have capacities in the range 600–3,000m^3. This type of plastic collar is delivered in sections which have to be heat welded on site. Manufacturer's prices normally include delivery and assembly.

Flexible, multi-sided cages   The most versatile flotation gear designed for large-scale operation, which can be used in exposed waters as well as more sheltered sites, consists of flexible sections of inflated, armoured hose linked at the ends to form polygons. The jointing sections can be angled to form rectangles, hexagons or octagons. Each section has bolted on vertical stanchions which carry the nets to prevent the fish leaping out of the cage, as well as supporting the main cage net.

The 16m sections made up as a hexagon cage, have a capacity of 6,500m^3. A larger octagonal cage has a capacity of 12,500m^3. The average density of Atlantic salmon which can safely be held in floating cages is 10kg per m^3. The larger 'high seas' cage can at times hold between 100 and 150 tonnes of salmon and the security of the whole operation rests on the strength of the cage nets.

The 'high seas' cage has the great advantage that it is designed to withstand the force of wind and wave when anchored off shore. Servicing these cages on open waters, particularly in rough weather, requires the use of a large, sea-going work boat and special equipment for feeding and handling the fish.

Nets and net mesh   Nets used for floating fish cages or enclosures should be woven of a strong, artificial fibre which is not affected by salt water. The netting should preferably be woven with a square mesh as opposed to the diamond mesh used for fishing nets, as it is hung as a flat wall not a loose curtain. Smaller mesh cages are best made up of knotless netting because it offers a greater open area for the passage of water, but cages to hold large fish are sometimes made of ordinary knotted netting as this is easier to repair.

The mesh of nets and the twine size used depend upon the size of fish to be kept in the cages. The largest mesh capable of preventing the escape of fish is not only the cheapest but also offers the least barrier to the interchange of water. The possibility of the fish either being gilled in the net or caught by their teeth is a limiting factor which must be taken into consideration in determining mesh size. The minimum mesh for salmonids at the stage they will tolerate

130

average marine salinity is approximately 12mm measured knot-to-knot.

Many different methods have been devised for mounting cage nets. The first consideration is to enclose the maximum volume of water within the depth of hanging net. Salmonids in floating cages tend to swim round the enclosed area. Their turning circle is a factor of their size. Large fish in square or rectangular cages can waste the water space in the corners. There is also a risk of abrasion causing skin damage and disease when the turning fish brush against the flat sides of the net. This is an important reason for using multi-sided or round flotation collars, particularly for cages holding large fish.

The nets supported by either round or multi-sided collars are in fact round as they hang. They can either be made up as a cup, or with vertical sides and a flat bottom. The cup is less expensive, and rather easier to handle and keep clean, but the flat-bottomed net holds more fish. All mounting points have to be designed to avoid abrasion and keep the net from rubbing on floats, walkways or linkages.

The vertical stanchions mounted on a flotation collar above the water carry a ring or net round the cage to prevent fish from leaping out. This net need not be so strong as the cage net but should be of a mesh small enough to prevent the fish from catching in it if they strike it when jumping. The net should stand $1-1\cdot5$m above the water.

Sea-nets must all be treated with an anti-fouling dip or paint. Most anti-fouling material contains copper salts or other poisons intended to prevent the attachment of marine plants and animals. These can also be poisonous to fish and endanger the environment. Nets mounted just below or above the surface of the water will deteriorate in sunlight and have to be renewed fairly frequently. They should be made with a separate detachable skirt and the twine should be dyed a dark colour. There is no evidence that the colour of cage nets has any effect on the fish enclosed but generally dark colours or black are preferred.

Mechanized cage systems   Giant cage flotation collars incorporating food stores, computer controlled fish feeding and hydraulic gear for handling nets and fish, operated by a built-in source of power, have been designed and tried out in prototype. Some systems involve vast rafts with multiple net pens, onboard facilities for fish processing, and living quarters for staff. It has been suggested that redundant oil tankers or container ships could be adapted as all-weather fish farms anchored in deep water. These ideas remain to be proven in practice as economically viable for the commercial production of farmed salmon.

Commercial salmon cages   *'Viking' cage system*   These are in the form of modules joined to form rectangular cages and are produced in Norway. They are made-up using wide metal framed and surfaced pontoons supported above the water on pairs of plastic flotation blocks. The paired blocks at intervals along each side of the pontoons act like

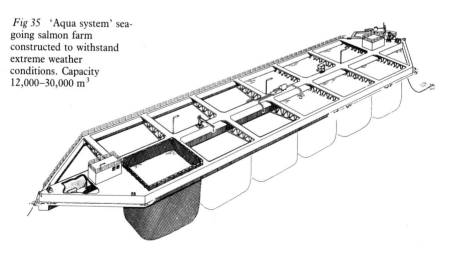

Fig 35 'Aqua system' sea-going salmon farm constructed to withstand extreme weather conditions. Capacity 12,000–30,000 m^3

catamarans and give added stability. Frames and handrails are of galvanized piping, and the walkway decking is made of perforated, galvanized steel. The wide pontoons provide a good working platform and enclose a surface area of 144m^2. The volume of the cage depends on the depth of the net.

'Atlantic' cage collars These consist of straight sections of twin plastic pipes, joined together at intervals by plastic plates, and at the ends by short, angled sections, to form octagons or squares. The inner pipe is filled with polyurethane foam. The sections have welded-on vertical pipes supporting a handrail which also carries a netting fish fence to prevent the salmon from jumping out. The standard enclosed areas of the square cages are 100, 144, 225 and 306m^2. The

Fig 36 'Atlantic' plastic cage collar. Straight sections (A) are joined by short angled sections (B), either 135° to form octagonal cages or 90° for square cages

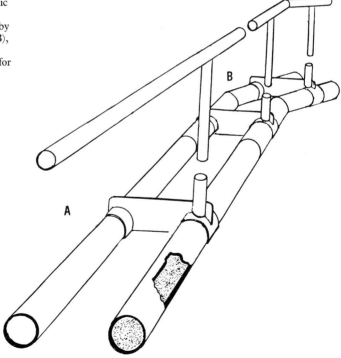

133

octagonal cages enclose surface areas of approximately 121, 198, 272 and 371m^2. The volume of the cage depends on the depth of net. About 75% of the theoretical volume is available to the fish.

'Polarcirkel' cages   These are round flotation collars manufactured in Norway made up from two concentric circles of plastic piping, heat-welded on site. The inner pipe is filled with polyurethane foam and carries vertical pipes welded-on at intervals which support a handrail carrying the fish fence round the cage. Cages range from 40–70m in circumference. The 60 and 70m sizes can be fitted with a wide walkway across the middle of the cage.

'Bridgestone high-seas' cages   These very large 'super' cages are fully flexible along the length of each section. The sections, which are made of inflated, armoured

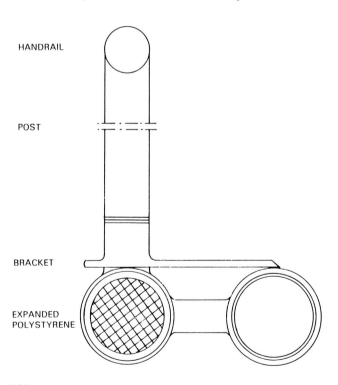

Fig 37 Section of 'polarcirkel' round plastic pipe cage collar

HANDRAIL

POST

BRACKET

EXPANDED POLYSTYRENE

134

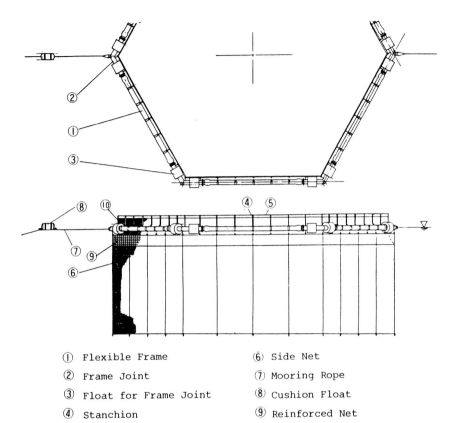

① Flexible Frame   ⑥ Side Net
② Frame Joint   ⑦ Mooring Rope
③ Float for Frame Joint   ⑧ Cushion Float
④ Stanchion   ⑨ Reinforced Net
⑤ Handrail Rope   ⑩ Fence Net

Fig 38 'High-seas' salmon cage—6,650 m³ capacity. Flexible sections absorb wave energy on open ocean sites

tubing, joined to form six or eight-sided cages, conform to the shape of the waves, absorbing their energy, even under storm conditions. Marine anchorages can be safely made in open waters. The large volume of the cages allows more space for exercise and provides a better environment for growing salmon. The 'super' cage is particularly suitable for 'one-shot' stocking in which big smolt, put to sea in April, can be harvested the following November. The standard size cages have enclosed surface areas of 665m² and 1250m² supporting 10m deep nets.

135

10 Sea farm management

The work on a saltwater farm is divided between
engineering maintenance and fish husbandry. The same
people very often have to do both jobs and they can
only be properly carried out by people who have had
thorough practical training in the specific operations
they are expected to perform. Many of these are
difficult and some can be dangerous, particularly when
working from boats or floating walkways on off-shore
cages.

Husbandry

Basic husbandry is much the same for fish fattened in
shore tanks, enclosures or in cages, but the jobs have
to be carried out in different ways.

*Grading Atlantic
salmon*

Growing fish do better if they are handled as little as
possible. Intermediate grading may be of no real
benefit and can only be easily carried out in shore
tanks. It is an awkward task with salmon in cages
moored in flotillas to stable walkways or to pontoons
connected to the shore, and very difficult in large cages
anchored separately. Grading is not practicable with
fish in large enclosures.

136

Grilse selection   Some farmers regularly grade in March/April. Very
experienced husbandry people can recognize the fish
that are going to mature as grilse. They are generally
slimmer and have a more deeply forked tail than fish of
the same age which will take a further year in the sea
before maturing as salmon. Grilse grading is too
difficult to be worth while when the fish are in very
large cages and impossible when they are in enclosures.
No grading is normally attempted and the race of
salmon selected for stocking should have a very low
'grilsing' rate (small proportion of fish which mature as
grilse after only one year in the sea). The age of the
fish in large cages or enclosures is of no importance.
All that matters is that they should reach the required
market size as quickly as possible and be in good
condition when they are slaughtered.

Feeding methods   Fish in shore tanks can be fed either moist or dry
pellets by automatic gravity or compressed-air feeders.
The same methods can be used for cages moored
inshore or in flotillas. Separately moored cages of the
'polarcirkel' type, with a wide central walkway can also
be fitted with automatic feeders. Fish in large cages,
either round or polygonal, are more difficult to feed.
Some sea-farmers feed by hand, particularly if they are
using moist pellets. The water 'cannon' feeder
described for use with the 'high-seas' cage can be
employed for the largest cages, or a portable
compressed-air feeder can be mounted on a work-boat
or pontoon. Compressed-air feeders need an engine and
a compressor, but more than one gun-feeder can be
'fired' from the same air tank. The 'gun' consists of a
barrel into which pellets are fed from a hopper and
then fired out at intervals by a blast of compressed air.
The fish in enclosures are fed by delivering the food
from the shore through a water pipe.

Treatment   Chemical baths and dips for bacterial diseases or
parasite control are given to fish in shore tanks by

137

careful control of degree of dilution and immersion time. It is also a simple matter to provide aeration if this is needed during or after treatment. Methods for treating the fish in cages become progressively more difficult and less effective as the cages increase in size. Fish in cages moored to stable walkways can be transferred to enclosures of plastic sheeting or tanks, where controlled treatment can be given without much difficulty. Fish in large, separately anchored cages have either to be treated in the main cage by raising the cage net and surrounding the shallow area with a plastic curtain, or by transfer to smaller service cages, moored alongside the main cage. Either method is awkward and laborious, even in the calmest of weather, and it can be particularly difficult to control the strength of dosages. The generally lower densities at which the fish are kept in large cages goes some way towards avoiding epizootic outbreaks of disease. Treatment of any kind is not practicable in enclosures, where the fish are on free range from start to finish, and diseases can only be dealt with by medication.

Capture and pre-market selection
Fish in shore tanks and cages moored to stable walkways can be easily captured. Undersized fish or fish unsuitable for sale are sorted and transferred by plastic tube or chute to other tanks and cages. Fish in large net cages have to be captured by reducing the area of the swimming space. Separate service cages moored alongside may have to be used to retain the fish which are to be returned alive to the main cage. The problem of re-capturing fish in large enclosures is solved by using a brailing net on a derrick to surround the area in which the fish are accustomed to feed.

Fish slaughter
Many ways of humane fish killing have been tried. The most satisfactory method is probably to release carbon dioxide (CO_2) into the water through a diffuser from a commercial cylinder. The fish in cages have to be

transferred to tanks or confined to a small part of the cage, enclosed by plastic sheeting.

The Norwegians use a method which although it may seem cruel, causes the fish very little apparent distress. A knife is inserted behind the head which severs the main blood vessels. This is done with lightning speed and the fish are then returned to the water, where they rapidly weaken and die through loss of blood. Norwegian fisherman often bleed the wild salmon they catch as this is thought to improve the quality of the flesh.

Supervision ashore and afloat

The majority of losses resulting from failure of water supply and equipment on shore, or flotation collars and nets in the sea, are due to human error. Salmon eggs, fry, parr and smolt are much too valuable to risk losses which can cripple future salmon production. The water supply to shore tanks must be entirely secure in terms of quality, quantity and temperature, at all times of the year, and attention must be paid to the risk of failure not only in hot, summer weather, but from low flows in winter during an exceptional period of prolonged frost.

Smolt reared in floating cages are not at risk from water shortage and generally not from pollution, but surface temperatures can get dangerously high, or reach a level when feeding should stop. The fish are difficult to see in the deep-water and the cages should be raised to check the condition of the fish and to inspect the mesh for damage or overgrowth with algae. Anchorages and moorings in freshwater lakes should be inspected by diving at the beginning of each rearing season, and securely buoyed if the flotation collars are moved or brought ashore.

The salmon in sea-cages are not only worth very large sums of money, but represent cash flow that depends upon each year class of fish being replaced by the next, without interruption. Cage-nets and ropes

139

must be frequently raised and inspected. Large cage-nets have to be checked by diving. If there is any sign of serious abrasion or other damage to the mesh, the net should be either replaced or repaired *in situ*. A diver can quite easily sew on a panel of net as a temporary repair. Anchors, anchor chains and submerged mooring ropes should be checked at regular intervals, and when high winds and heavy seas die down.

Predation   Flotillas, or large nets moored separately, should be completely protected by a predator net suspended in a curtain from the walkway, or round the outside of the flotation collar. Predator nets can also be mounted away from the cages on separate floats. The mesh should be small enough to exclude predatory sea-birds such as cormorants or fishing ducks. Although the birds cannot get at the fish they cause injuries by stabbing them through the cage mesh.

Attacks by seals can cause serious losses if they manage to reach the main cage. Some salmon farmers go to the extent of extending the protecting net under the whole cage or flotilla of cages.

Top nets are needed covering the surface of each cage to exclude fishing birds. Mounting these nets above the water presents no problems in smaller rectangular or polygonal cages, but nets over very large cages may have to be supported on separate floats such as large blocks of polystyrene foam.

Security   A range of electronic devices are available designed to protect both shore and sea farms, but human supervision is the most effective. Inaccessibility provides the best protection from interference or theft. Fry, parr and smolt are normally only at risk from sabotage or vandalism. Lighting the shore-site at night is a useful deterrent.

Salmon in saltwater tanks or cages that can be

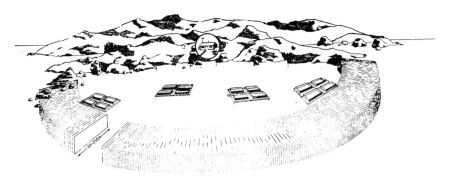

reached by walkways from the shore are open to the risk of deliberate damage and also from theft, but the value of the fish makes constant human supervision the only sound, economic safeguard. Offshore flotillas of cages or single, large units are at some risk from theft, and fish have been stolen from the cages, but this is usually a well-organized act carried out from a fishing boat coming alongside the cages at night.

Electronic alarms can be activated by closed-circuit TV, infra-red sensors or by changes in the loading on walkways. The best protection that can be given to sea-cages is by short-range radar, strategically positioned on shore to scan the whole marine site area. The radar can be designed to activate both a central alarm and individual alarms carried by staff.

Brood stock Salmon of a particular genetic strain may be grown on to maturity as brood stock, either for use at a freshwater shore site producing smolt, or on behalf of the specialist smolt farmer that supplies the sea farm. The fish do not require any special attention, apart from the feed given when the gonads are ripening, but they should preferably be kept at a reduced density compared to market fish. It can be profitable to 'mend' female kelts and this presents no problems, provided the fish have been carefully handled and anaesthetized before being stripped of their eggs. The previous

141

spawners grow well and produce large, healthy eggs. It is not worthwhile trying to restore male kelts.

Intrusion of other species of fish
A problem which arises for salmon farmers using enclosures or cages for their fish in the sea is the uninvited entry of unwanted sea-fish. Some of the fish species involved, notably saithe (*Pollachius virens*) and pollack (*Pollachius pollachius*) get into the cages as young fish, and grow to large size on the food supplied for the legitimate tenants. They can be a nuisance as well as consuming expensive food, particularly if valuable stocks of brood fish are being held in sea-cages which they have invaded.

The only way to remove unwanted species is to shift all the fish into a small working cage and extract the interlopers. Salmon can be badly stressed if hunted round a small cage. The correct procedure is to net-out the unwanted species, not try to net the salmon, although they may be greatly outnumbered.

Salmon welfare
All good stockmen should know when their animals are doing well. Fish are much more difficult to assess than terrestrial beasts. They inhabit the water from which we all originated. We are beginning to re-discover some knowledge of life in the sea through the eyes of scuba divers, but it will be a long time before we properly understand the needs of fish.

Salmon are animals genetically programmed to spend most of their lives swimming freely through the oceans. We now confine them in tanks or cages in close proximity and frequent physical contact with thousands of others. In the open seas they would probably never have come as close to any other fish of their own kind before returning to spawn.

Most of the more dangerous diseases and parasitic infestations are density dependent. Salmon farmers are continually staving-off disaster in the battle against the onset of new diseases or increasing resistance of the old ones. We may have to think again about the environment we provide for the fish. Extra large cages of the 'high-seas' type, where the growing salmon have more space to exercise, may be a step in the right direction, but health and freedom from disease really depends upon a better understanding of the well-being of the fish.

11 Fish feeding and nutrition

The food originally provided for members of the salmon family, hatched to supplement wild populations or grown-on to re-stock angling waters, was either the same or resembled as closely as possible what the fish took naturally. The increasing domestication of salmonids as a result of their being farmed for the table market has led to the investigation of a whole new range of nutritional needs. Carnivorous fish require a diet very different from the mammals and birds that have been farmed in the past, in terms of proteins, fats, carbohydrates, vitamins and minerals.

The formulation of diets and their development as a part of the animal feed industry has taken place comparatively recently, but dry crumb or pelleted fish foods, first produced in the USA, are now a commonplace item which can be prepared in any mill with a pelleting machine that uses a satisfactory formula and the right ingredients. Good quality dry fish feeds are now completely successful for all stages of the freshwater culture of salmonids. They are not so successful by themselves for feeding salmon and trout at sea, although a milled meal with added vitamins is now generally regarded as an essential constituent of the moist diets used for sea-feeding.

The protein in a fish diet contains the essential amino acids which have to be present in the correct proportions to promote the vital physiological functions of the body. Fats of low melting point, usually fish oils, are the main sources of energy. The essential polyunsaturated fatty acids are linoleic, and arachidonic. High-fat diets are essential for sea-going salmonids in salt water and for the production of S1 smolt in fresh water. Some carbohydrates, in small quantities, can be digested by salmonids to provide an additional source of energy.

A good deal is now known of the vitamin needs of salmonids and vitamins are added to all dry diets and to the meals combined with wet food mixtures. Vitamin deficiencies cause symptons that can be mistaken for those produced by some bacterial and virus diseases. The calcium and phosphorous needed for bone formation must form part of the diet. Traces of other minerals are needed to promote catalytic processes in the body.

Animal protein which is the main constituent of all good quality fish diets is expensive. Other proteins such as soya and feather meal are also used in commercial food formulation. The single-cell proteins have proved a satisfactory alternative, particularly the variety derived from methanol by the action of methylophilis bacteria. In the longer term, algal protein may prove to be the solution to feeding domestic carnivorous fish. This rational progression has already taken place in husbandry on land where plant

Table 3 Basic constituents of commercial dry salmon feeds

| Materials | (*Percentage of total weight excluding ash and fibre*) | |
	Juvenile fish in fresh water	*Growers in salt water*
Protein	50–55	45
Fats	17	16–18
Carbohydrate	11	18
Water	8–9	8–9

crops are grown, harvested and processed to feed terrestrial farm animals.

Dry feeds   Commercial fish food manufacturers have their own formulations, usually based to some extent on local availability of ingredients. *Table 3* shows the basic proportions of the constituents in high quality dry feeds made up for feeding juvenile salmon in fresh water and for growing-on the fish in salt water. Vitamins and minerals are added to the dry food mixtures.

Special, fat-reduced formulations are made up for feeding brood fish. Feeds containing carotenoids can be obtained to pigment the flesh of fish prior to slaughter. Medical feeds for the treatment of bacterial diseases are also available, either directly from the feed manufacturers in some countries or under veterinary prescription in others. Binder meals for mixing with wet feeds contain a cellulose or alginate binding agent and additional vitamins.

Wet or dry feeds   There is no argument about the economy and general superiority of dry feeds over wet feeds in fresh water. They are superior in most respects not least of which is reduced pollution but wet feeds may be more economic where trash fish is available. The only problem with dry feeds used for young fish is the formation of dust during transportation and storage. This can enter the gills of the fish and promote bacterial gill disease. The best high-fat diets avoid the problem because they are slightly tacky and do not form dust.

The controversy over the use of all dry feeds for feeding fish in salt water is whether or not their exclusive use causes osmotic stress. The concentration of salts in the body of anadromous fish in fresh or salt water must remain within certain prescribed limits in

146

order for them to survive. In the sea, fish which have to drink sea water must get rid of the excess salts. This is an active process demanding the use of energy.

Essential amino acids in diets

The following amino acids are thought to be essential in the protein content of salmon diets. The quantities expressed as a percentage of the total protein in the food are as follows:

Argenine	6·0%	Methionine	1·3%
Cystine	2·5%	Phenylaline	3·0%
Histidine	1·7%	Threonine	2·3%
Isoleucine	2·5%	Tryptophane	0·5%
Leucine	3·9%	Tyrosine	2·0%
Lysine	5·0%	Valine	3·2%

There are often considerable differences in the amino acid make-up of the various proteins used in the commercial preparation of dry feeds or wet food supplements. An analysis of the amino acids in a typical sample of high-quality dry feed, expressed as a percentage of the total protein in the diet, is approximately as follows:

Alanine	3·2%	Lysine	3·4%
Argenine	2·7%	Methionine	1·3%
Aspartic acid	4·2%	Phenylaline	2·1%
Glutamic acid	6·8%	Proline	2·7%
Glycine	3·0%	Serine	2·1%
Histidine	1·6%	Threonine	2·0%
Isoleucine	1·6%	Tyrosine	1·4%
Leucine	4·0%	Valine	2·7%

Vitamins in fish diets

The quantities of the different essential vitamins that are added to dry food mixtures depend on the amounts naturally present and retained in the ingredients after processing and storage.

The approximate requirements for water-soluble

147

vitamins (in mg per kg dry weight of food) for
salmonids are as follows:

Thiamin (B1)	10–12	Folic acid	6–10
Riboflavine (B2)	20–30	Inositol	200–300
Pyridoxine (B6)	10–15	Choline	500–600
Pantothenic acid	40–50	Cyanobalamine	trace
Nicotinic acid	120–150	(B12)	
Biotin	1–1·2	Ascorbic acid (C)	150–450

Salmonids also need the fat-soluble vitamins (A, D,
E and K). These differ from the water-soluble vitamins
because they can accumulate in the body and cause
vitamin poisoning or hypervitaminosis. The quantities
of these vitamins given in a dry food mixture or
supplement must therefore be adjusted to the quantity
and the type of fats in the original mixture. The
following are average amounts per kg of fish food.

Vitamin A	8–10,000 iu
Vitamin D	1,000 iu
Vitamin E	125 iu
Vitamin K3	15–20mg

Minerals   Wild salmon obtain the essential minerals they need for
healthy growth from ingesting the water in which they
live and from the tissues of their prey. Minerals are
added to commercial dry feeds and wet food
supplements. Calcium and phosphorus must be
available in unrestricted quantities. Dry fish meal
contains about 26·0g of calcium and 21·0g of
phosphorus per kg which is ample for the diet. If one
of the single-cell proteins and soya is the main source
of protein in the diet there will be too little calcium
and phosphorus and more will have to added as
calcium carbonate and di-calcium phosphate. The
minerals added to proprietory mixtures are shown in
the example of a commercial diet.

Fats   High-fat diets are now considered essential to promote

148

quick growth in fresh water, early smolting and on-growing in the sea. The amount of supplementary fats which may have to be added in the form of fish oils depends on the fats in the dry constituents of the mixture. The make-up of various dry meals is shown in *Table 4*.

It can be seen from *Table 4* that in dry feeds where fish meal and soya are the main ingredients, assuming that the proportion is constant and bearing in mind that the fat content of different soya meals can vary considerably, the amount of fat in the initial mixture will depend on the original fat content of the industrial fish in the fish meal. For example, if capelin (*Mallotus villosus*) in pre-spawning condition were used to make the fish meal and the proportion of the other ingredients remained the same, the fat content of the pellets would be four to five times greater than if the fish meal had been made from white fish.

The fat content of a dry food mixture will also be

Table 4 Raw materials that may be used in dry pellets and meals

Nutrient components (crude protein 40–55%)	Minerals (not more than 3% of total wt.)
Fish meal	Calcium carbonate
Soya meal	Di-calcium phosphate
Wheat meal	Manganese sulphate
Cane molasses	Magnesium sulphate
Fish oils	Iron sulphate
Soya oils	Iron carbonate
Zinc sulphate	Iron oxide
Lecithin	Copper carbonate
Vitamins	Zinc sulphate
Anti-oxidants	Zinc oxide
	Potassium iodide
	Sodium carbonate
	Sodium chloride
	Cobalt sulphate
	Cobalt carbonate

(all these components are not used in every formula and others may be substituted by manufacturers)

149

raised considerably by using a high-fat soya or by increasing the proportion of dry milk solids. The utilization of fats in salmonid diets is improved by the addition of lecithin to the dry food mixture. Final adjustment of the fat content by the addition of fish oils is only possible within certain limits as the mixture can become too moist to form a good pellet.

If the components of salmonid diets lack the essential fatty sources of energy for balanced nutrition this is, in practice, made up by adding digestible oils directly to the other constituents. The presence of linolenic acid is essential and fish oils contain an average of 30% compared to only 6–8% in soya oil. Fish oils are therefore the preferred source of supplementary unsaturated fatty acids.

Preservation of fats in salmonid diets

Rancidity in the oils used to formulate dry feeds or as additives in wet food mixtures can result in serious, permanent damage to the digestive system of the fish, even if fed for only a short time. The fats used in diets should be as fresh as possible and treated with an anti-oxidant solution which should be incorporated in fish oils at the rate of 2g per litre. Vitamin C is also a useful anti-oxidant.

The vitamin E (alpha-tocopherol) content of high-fat diets is of importance in hindering the oxidation of fats during storage. Approximately 0·4g of vitamin E should be added per 100g of fat in each kg of the mixed food (*ie* a food mixture with 10% fat content will require 0·4g vitamin E per kg and pro rata).

Manufactured dry feeds

The constituents of commercial pellets and meals vary widely, due mainly to the availability and cost of raw materials. The essential factors in a good dry feed are that its composition should be consistent and its manufacture carefully controlled so that it is fully up to the manufacturer's specification. Basic raw materials commonly used to manufacture dry, pelleted fish foods are given in *Table 4*.

150

Adequate vitamins are normally present in commercial dry feeds. Thiamin may be added as an additional precaution against thiaminase if the pellets have been stored for several weeks and it is suspected that the main source of crude protein was fish meal derived from fish of the herring family.

Energy in dry and wet feeds

The energy consumption of fish is measured in kilo-calories (kcal), or mega-calories (mcal) (1mcal = 1000kcal). The total available energy in fish foods is measured in usable or metabolizable energy (ME), which includes the energy remaining in the body's waste products. The sources of energy in fish feeds are proteins, fats and carbohydrates. The digestibility of a particular diet will vary according to its components.

In dry feeds the average digestibility of protein in pellets and meals is about 80%. That is to say the fish can digest about 80g out of every 100g of protein they eat. The digestibility of the fat content in the form of fish oil is about 85%.

Salmon can only tolerate about 12% by weight of digestible carbohydrate in any diet. More carbohydrate is often included in food mixtures but this is safe if it is indigestible and serves no useful purpose except as a filler.

The metabolizable energy yields for the main groups of components in the average dry feed are as follows:

Material	Kcal ME per g
Protein	3·5
Fats	8·0
Carbohydrate	2·3

Fresh wet feeds

In practice sea farmers have found an improved economic return by feeding minced raw or deep-frozen industrial fish and crustacea mixed with a vitaminized

151

meal and binding agent to form a thick porridge.

The average energy content of diets composed of minced industrial fish or fish offal, with the addition of a source of carbohydrate in the binding agent, are as follows:

Material	kcal ME per g
Protein	3·9
Fats	8·0
Carbohydrate	1·6

Food value   The nutritional quality of commercial dry feeds is nearly always given as a conversion factor or ratio. This simply means the amount of food in kg it takes to produce 1kg of fish (usually ignoring the not inconsiderable amount which is wasted). It is not this simple ratio which decides the economic value of a fish food but the metabolizable energy (ME) which it provides to the fish. This is why it is often essential to incorporate comparatively large amounts of digestible fats in the form of fish oil in the diets of salmonids as these are their main source of ME. A low energy content in fish food is the explanation for poor growth in fish although they have been fed what is apparently a high-protein diet.

Wet fish food mixtures   *Table 5* shows how the protein and fat components of different wet fish food sources vary due to the fish species available and the time of year when they are caught. The fat content has the greatest range and this produces a wide range in the ME per kg of raw food. This can vary from about 1000kcal to more than 1900kcal per kg. This means that at least double the quantity of low-energy food will be needed to produce the same growth as from a high-energy food. The main source of energy is the metabolizable fat in the food or food mixture. It is this which may have to be

152

Table 5 Marine animal tissue in wet diets

Species and feed	Dry wt.	(g and ME kcal per kg of feed) Protein	Fat	Kcal	Fat energy %
Capelin (BF)	290	130	140	1630	70
(*Mallotus villosus*) (AF)	200	130	40	830	58
Sprat or brisling (BF)	310	160	130	1650	63
(*Sprattus sprattus*) (AF)	285	160	105	1480	53
Other clupeids (Herring family)	300	160	130	1650	62
Herring offal	230	150	40	1000	50
Blue whiting (BF) (*Micromesistius*	260	165	75	1243	48
poutassou*) (AF)	240	165	35	950	30
Saithe (whole) (*Pollachius virens*)	250	175	50	1120	40
(offal)	235	165	30	990	30
Norway pout (BF) (*Boreogadus esmarkii*)	300	165	120	1608	60
(AF)	220	165	20	820	20
Sandeels (*Ammodytes* spp)	270	180	60	1320	52
White fish offal	200	160	25	965	24

BF = Before Spawning
AF = After Spawning

maintained at the desired level by the addition of fish oils.

The total fat requirement of salmon is thought to depend on water temperature ranging from about 7–8% of the diet in winter to 18–20% in summer.

The use of any particular fish or shellfish as the main component of the diet of salmonids will depend upon availability and price. Given that there is some element of choice, the fat content can be adjusted either by choosing a fatty fish or by the addition of the right

153

amount of oil to bring up the calorific value to the desired level.

The digestible material in food mixtures (DM) to form a satisfactory wet diet should result in a balance between the energy (kcal ME) derived from protein and from fats. If this cannot be achieved from the available raw materials the fat energy source must be artificially increased by adding fish oil. The final mixture should have a total weight to dry-weight ratio of about 3 : 1 to give the right consistency. Moist, rather than wet foods can be made by increasing the dry content to include more nutrients in addition to the binding agent, but mixtures of this kind fall more into the category of moist pellets.

High-quality raw materials usually contain adequate amounts of the essential vitamins but when fat fish of the herring family are used as the main source of protein, thiamin should be added in the proportion of 0·2g per kg of total weight.

Wet feeds, raw materials and costs   The factors to be taken into consideration in producing a wet food mixture for on-growing salmonids in the sea are total protein, total fat, ME balance between protein and fat energy sources and cost. The following diets illustrate the results in terms of nutrient value of different mixtures.

The first mixture has added shrimp waste as a source of astaxanthin to pigment the flesh red. An artificial substitute could be used for this purpose. The ash content is not included in the figures.

This diet is rather low in calories as it only contains 1,163kcal per kg. It has adequate protein but too little fat. If 20g fish oil and 10g of lecithin per kg is added to the mixture this would increase the total calories to about 1400kcal per kg. It would also bring the fat energy level up to 59·1% which would then be in balance with the protein energy.

Diet 1 is basically a lean, low-fat mixture. It is better to produce a balance between the energy sources

154

Table 6 Diet 1.

| Feed Mix | Ingredients per 100kg of wet food mixture | | | | | |
	Total wt kg	Dry wt kg	Protein kg	Fat kg	Carbohydrate kg	ME mcal
Blue whiting	40	10·4	6·6	3·0	—	49·7
White fish offal	40	8·9	6·4	1·6	—	37·8
Prawn waste	10	2·4	1·3	0·2	—	6·8
Vitamin meal	10	8·8	0·9	0·3	7·0	22·0
Totals	100	30·5	15·2	5·1	7·0	116·3
Nutrient %	—	—	49·8	16·7	22·9	—
Energy %	—	—	50·0	35·1	31·5	—

Table 7 Diet 2.

| Feed Mix | Ingredients per 100kg of wet food mixture | | | | | |
	Total Wt kg	Dry Wt kg	Protein kg	Fat kg	Carbohydrate kg	ME mcal
Industrial clupeid	45	13·5	7·2	5·9	—	74·3
Saithe offal	25	5·9	4·1	0·8	—	24·8
White fish offal	20	4·0	3·2	0·5	—	19·3
Vitamin meal (added thiamin)	10	8·0	0·9	0·3	7·0	22·0
Totals	100	32·2	15·4	7·5	7·0	140·4
Nutrient %	—	—	47·8	23·3	21·7	—
Energy %	—	—	42·8	42·7	8·0	—

without having to add extra fats.

Diet 2 a much better mixture than Diet 1. The ME kcal is 1,404 per kg. There is adequate protein and a good balance between the protein and fat energy.

It is quite possible to use only a combination of locally available industrial fish or fish offal provided that the protein and fat contents are known. It is also possible to add up to 30–35% of the total weight of the mixture in the form of dry vitamin meal, based on good-quality fish meal. The mixture will still remain sufficiently liquid to be easily distributed.

155

Binding agents for wet and moist food mixtures   The binder most commonly used which gives good results is methyl-cellulose (carboxymethylcellulose = CMC or hydroxypropylmethylcellulose = HPMC). This substance quickly absorbs water to form a stable, water-soluble gel. Approximately 1% may be added directly to the ingredients of the mixture but for wet feeds the binding agent usually forms part of the vitaminized meal which contributes about 10% by weight of whole feed.

Alginates may be used as binding agents in the preparation of moist pellets. Their binding efficiency depends upon their degree of polymerization in the presence of free calcium ions. A source of calcium ion has to be provided in the food mixture. Approximately 30mg of calcium ion is needed per g of alginate. If calcium hydrogen phosphate ($CaHPO_42H_2O$) is used as a source of ion, 130mg should be added per g of alginate. A small quantity of NaCl may also be added, depending on the salt content of the ingredients in the food mixture and whether the feed is being given in sea water.

Moist pellets   The moisture content of wet feeds usually averages about 70%. Moist pellets have a moisture content of 20–50% of their total weight. The pellet mixture consists of the following basic components:

— fresh industrial fish and fish offal;
— fish and other protein meals, including some carbohydrate;
— fat in the form of fish oils and milk concentrates;
— vitamins, minerals, anti-oxidant and binding agent.

The advantage of moist pellets is that they do not disintegrate to the same extent as wet feeds and can be autofed. It is a comparatively straightforward operation to manufacture moist pellets using a modified industrial mincing machine, fitted with a revolving blade. The blade cuts off the 'worms' of pre-mixed

156

food in short lengths as they come out of the perforations on the face-plate of the mincer.

Pigmentation of salmonid tissues The pink or red flesh colour is a characteristic of sea-going members of the salmon family that should be produced in farm fish. The natural pigment is a carotenoid called astaxanthin which occurs in marine animals such as prawns, shrimps, krill and certain plankton copepods. The fact that a fish species feeds on these animals does not necessarily mean that it has pigmented flesh. The red or pink muscle colour depends on the ability of the cells in particular fish species to retain the carotenoid pigment. Some salmonids are better at this than others and will develop deep red flesh by ingesting a smaller amount of carotenoid for a shorter period than others.

Natural sources of astaxanthin used by European fish farmers to form part of fresh-food mixtures are as follows:

Species of crustacea	Astaxanthin mg per kg
Prawn waste (*Pandalus borealis*)	97–128
Plankton red copepod	76–84
Krill	73–98

The artificial carotenoid, canthaxanthin, is also used to pigment salmonid muscle. It produces an attractive colour more quickly than most natural sources of astaxanthin. A concentration in the tissue of 4mg per kg of carotenoid pigment is needed to produce pigmentation in salmonid muscle.

Salmon flesh will not attain a satisfactory colour until the fish have been fed with the carotenoid for about six months prior to slaughter. It is possible to reduce the time taken to pigment the flesh by increasing the dose. A concentration of 450mg per kg in the food is needed to produce the deep red colour that is naturally attained by some species of Pacific salmon.

157

A relatively new source of dietary pigments for salmonids is the red yeast *Phaffia rhodozyma*. This contains astaxanthin as its main carotenoid pigment. The yeast can contain 500–800μg of astaxanthin per g, dependent on the growth conditions and the strain used. This concentration is about ten times higher than that in prawn or shrimp waste. *Phaffia rhodozyma* has the same properties as ordinary brewer's yeast and can be used as a supplementary nutrient to salmonid diets as well as a source of red pigment.

Mechanical fish feeding with wet food mixtures

Machines developed for dispensing wet feed consist of a large container with an air-tight lid, a compressor and an air receiver to create pressure in the food tank, which is connected to an articulated pipe with a trigger-operated mechanism to eject the food. Experiments made in Denmark have demonstrated that machine feeding with wet food produces at least as good growth in the fish as hand-feeding and is more economical. There is a considerable saving in man-power (it takes about one fifth the time to feed by machine that it does by hand). The fish also consume more of the food and there is less waste and consequently less pollution of the water or fouling of ponds or enclosures.

Saltwater farmers using separate off-shore cages have not made much use of machine feeding. It has been used for fish in cages anchored to the shore and in enclosures. The machine feeders and food pumps so far developed for wet food eject a single jet of food and the amount fed depends on the bore of the food-pipe. An improvement would be to devise an ejector-head with multiple openings that could deliver a number of smaller jets at the same time.

Automatic feeders for dry food

A wide variety of machine feeders capable of distributing dry feed are available from manufacturers of fish farming equipment. They have control

158

mechanisms capable of sensing light intensity and water temperature, as well as distributing the food at pre-determined intervals.

The use of high-fat dry feeds has caused problems with some automatic feeders as the tacky food tends to clog the mechanism. The most satisfactory of all types of automatic feeders for dry feed are those operated by compressed air. A blast of air drives a quantity of food from a pipe fixed to a hopper, out over the fish tank or enclosure. The source of compressed air can be either from small individual compressors or from a central unit through pipe lines. The delivery of food is controlled by sensory devices and time clocks. This system has been satisfactorily adapted to feed salmon at most stages of growth.

Fish feeding   The manufacturers of commercial dry pellet or crumb feeds for salmonids provide tables showing the amount of their feed required by different species at any given age or weight, according to water temperature; and the intervals at which they should be fed.

Sea farmers generally have their own ideas as to when and how their fish should be fed to give the best results. The basic principles are as follows:

— Salmonids cultured in sea water are usually given as much food as they will eat at the prevailing water temperature. The daily requirement varies with the kcal of the food as well as with the water temperature. A rough average of the daily food intake, allowing for waste, is about 7–9% of the body weight of the fish.
— Allowing for waste, about 4,500kcal ME is needed to produce 1kg of fish.
— The food must be properly distributed to a large surface area, but it should be directed towards the middle of cages or enclosures where there is not much tidal flow, in order to avoid waste.

159

Fish foods—raw materials and storage

1 Rancidity or putrefaction is particularly dangerous. The wet fish and fish oils used in wet diets or to make moist pellets, must be fresh. Any smell of ammonia is a first indication of decay.

2 Whole industrial fish or fish offal in cold-store should be held at below $40^\circ C$.

3 Freezing can cause deterioration in fats and vitamins in raw food materials. Fresh food should not be stored deep-frozen unless absolutely necessary.

4 Dry pellet foods should be stored for as short a time as possible, preferably in purpose built silos.

5 It is essential to make sure that dry components in food mixtures are up to specification, particularly vitamin meals and mineral supplements. Their contents should be checked periodically by an independent laboratory.

Deficiencies in salmonid diets

The most likely cause of poor growth is that the protein-fat content is out of balance, usually with too much protein and too little fat. The use of indigestible fats of high melting-point containing saturated fatty acids causes degeneration of the liver. Rancid fish oil can cause pancreatic damage with symptoms not unlike those produced by the virus disease known as IPN.

Serious digestive disorders, the effects of which are usually irreversible, are caused by feeding dry or wet diets containing putrid ingredients which have deteriorated in storage. Dry food which has become mouldy or smells musty can be actively toxic to fish.

The majority of dietary deficiencies are due to a lack of essential vitamins or minerals in the food mixture, caused either by deterioration or by partial destruction in preparation. Some of the following symptoms attributed to particular vitamin deficiencies have only been observed in the laboratory:

Thiamin (B1). Loss of appetite and impaired equilibrium; convulsive movements in advanced stages.

160

Riboflavine (B2). Loss of appetite; fish seek shade or darkness and swim deep; eye lenses may be clouded and eyes blood-shot; skin darkens.

Pyridoxine (B6). Loss of appetite; hyper-activity; rapid breathing and gasping; quivering of gill covers; fluid collects in the body cavity; anaemia; skin darkens; early rigor mortis.

Biotin (H). Loss of appetite; muscular atrophy and convulsive movements; skin darkens; intestinal lesions.

Nicotinic acid. Loss of appetite; jerky movements; fluid collects in stomach and intestine.

Pantothenic acid. Loss of appetite; gill filaments clogged with mucus; skin lesions.

Folic acid. Poor growth; sluggish movements; anaemia; skin darkens; fraying of fins.

Inositol. Poor growth; distended stomach.

Choline. Poor growth; fatty degeneration of the liver; bleeding in kidney and intestine.

Alpha-tocopherol (E). Poor growth; skin darkens; pancreatic damage which can be mistaken for IPN.

Ascorbic acid (C). Loss of appetite; lethargy; resting on the bottom. Symptoms may not appear for about twenty weeks of feeding. The normal recommended quantity in the diet should be increased if the fish are stressed in any way such as by transport, grading or handling for anaesthesia and external medication.

Fish food supplies Fish farmers in most developed countries have no problems in obtaining dry fish foods and some choice is usually available in terms of quality and price. Supplies of fresh or deep-frozen industrial fish for making up wet feeds or moist pellets for salmonids in salt water can be difficult and even impossible to obtain, at present, for some sea farmers in countries where there is little or no fishing for industrial fish.

There is no doubt that some of the most successful

salmon farmers are those who operate their own fishing boats and catch their own supplies of fresh food fish. This is a course of action strongly to be recommended as it makes the farmer independent of outside economic pressures and difficulties of supply. Surplus fish can be sold for fish meal or cold stored. Fishing can be on a co-operative basis between a group of sea farmers. Non-industrial species captured can be marketed for human consumption.

12 Disease control and hygiene

Disease is an abnormal and physically damaging
condition which can be caused by changes either inside
or outside the body of an animal. Diseases in fish may
result from any of the following conditions:

— bacterial or viral infection;
— infestation by internal or external parasites;
— environmental conditions such as lack of oxygen,
   entrained gases in the water or physical damage
   following skin abrasion or gill clogging;
— toxic algal blooms;
— deficiencies or toxins in the diet.

Much of the disease risk to cultured fish is the direct
result of their being held at farm densities, in
enclosures from which they cannot escape. Viruses and
bacteria known to be present in wild fish stocks only
occasionally cause epizootic outbreaks or large
casualties. The same applies to diseases resulting from
infestation by parasites and naturally adverse
conditions in the environment.

Some disease pathogens are present only in fresh
water, some in the sea and others in both fresh and salt
water. The sea-going salmonids are doubly at risk.
Diseases can be transferred from fresh to salt water

163

within the young fish, or the pathogenic effects of a disease which infected the fish while in fresh water may become apparent when they are stressed on removal to the sea.

Fish pathogens can be separated into two main groups. Those which are termed obligate are normally absent from water in which there are no diseased fish or carriers of disease. Many of the common bacterial and viral diseases belong to this group. The second group is termed facultative. These are pathogens which are naturally present in the water and may infect fish and cause symptoms of disease when they are stressed or there are physical changes in their environment, such as abnormal fluctuations in temperature or salinity.

The diseases specified are those most likely to be encountered by salmon farmers at some stage in fresh water or in the sea.

Viral diseases

Fish diseases caused by viruses seldom if ever respond to treatment, although secondary infections can sometimes be successfully treated. The only method of control is by isolation and destruction of infected stock. Samples of diseased fish must be submitted to a pathological laboratory for definite diagnosis, using live cell-cultures. Reporting the presence or suspected presence of a viral fish disease is a legal obligation in many countries. Some salmonids may resist infection or carry a virus disease, without showing symptoms, which is deadly in other species.

Infectious haematopoietic necrosis (IHN)

The disease is endemic in North America and Japan but has not yet been reported in Europe. The species mainly at risk are Pacific salmon. The virus can be carried in eggs, sperm or faeces shed into the water, or transmitted by feeding on infected fish tissue. Fish can become healthy carriers. It is a low-temperature disease and does not manifest itself above 15°C.

164

Symptoms Frenzied, erratic swimming (flashing); fry float upside-down, breathing rapidly prior to death; early signs are opaque faecal casts trailing from the vent and seen floating on the water or collected on outlet screens; areas of bleeding on the body surface at the base of the fins and vent; fluid in the body cavity; areas of bleeding in the wall of the body cavity and internal organs which are pale in colour; in alevins, signs of bleeding in the yolk sac which is distended with fluid.

Infectious pancreatic necrosis (IPN) The disease is endemic in North America, Europe and Japan. Infected fish can become healthy carriers. The virus can be transmitted through eggs or shed into the water with faeces. It is resistant to adverse conditions and probably remains active in a damp environment for some time outside the body of a host. Mortality is high among very young fish with a death rate of up to 85%. Survivors become carriers probably for life. It is not clear whether the disease can be transmitted in salt water, but otherwise healthy fish, carrying the disease, may develop symptoms when stressed by transfer from fresh water to the sea.

Symptoms Erratic swimming; corkscrew movements (flashing); fish sink to the bottom before death; internal bleeding.

Salmon farmers have to learn to live with this disease. Stocks appear to build up a degree of immunity and this has been confirmed by the inoculation of adult rainbow trout with live virus. Fish over six months old are naturally resistant to the disease.

It has been suggested that iodophors may reduce losses in salmonid fry when sprayed on dry food pellets or crumb at a rate of $1 \cdot 5 - 2 \cdot 0$g of undiluted commercial iodophor per kg of food fed for fifteen days. The 'treatment' may do no good but it will not do any harm as the fish would probably die in any case and the iodophor will at least act as a disinfectant.

165

Viral haemorrhagic septicaemia (VHS) The disease occurs on the mainland of Europe. The common mode of transmission is by live fish through the water or by wet, infected equipment. The virus is not transmitted in eggs. It is a cold water disease and outbreaks occur most frequently in winter, particularly when fish are stressed by grading or transportation. Fish of all ages are susceptible and the death rate in bad outbreaks can be up to 90–95%.

Symptoms Darkening of the skin; eyes bulging, with bleeding in or about the sockets; pale gills; fish become weak and lathargic; body cavity filled with clear or yellowish fluid; swollen and discoloured liver and kidney; flecks or blood and bleeding from the walls of the body cavity.

Bacterial diseases
Furunculosis The disease results from infection with the bacterium *Aeromonas salmonicida*. It is primarily a disease of Atlantic salmon and is endemic in wild as well as farm stocks. It can occur in fresh water and in the sea, and is the most common bacterial disease in farmed salmonids. Outbreaks are likely when the water temperature is 15–18°C or higher. The disease spreads by direct contact between the fish in tanks or cages or through the water. Fish can retain low concentrations of bacteria in their tissues and become carriers without showing clinical symptoms. Initial infection most commonly occurs during freshwater life when the young fish are grown in a water supply derived from a river holding wild salmon. In the sea the disease can be transferred over distances of up to several miles between cage farm sites by water movement or possibly marine fish.

Most serious outbreaks of furunculosis are the result of bad husbandry and overcrowding the fish in tanks or cages. The disease can be eradicated in sea sites if the cages can be allowed to lie fallow without fish for a period.

Smolt which have become infected by the bacterium during freshwater life are likely to develop clinical

166

symptoms of the disease when they are stressed by the transfer from fresh to salt water. Outbreaks commonly occur in post-smolt when sea temperatures start to rise during the early summer. The fish are particularly vulnerable in shallow, southern sites where the sea in coastal waters is likely to get very much warmer than in deeper, northern fjords or sea lochs.

Serious outbreaks of furunculosis are best avoided by good husbandry, avoiding stressing the fish and by careful handling at all stages of growth. The best salmon farmers have learned to live with and to control the disease although it is endemic and some incidence is inevitable.

Symptoms The disease develops after an incubation period of three to four days and salmon parr or young trout may die in large numbers without showing any symptoms other than a slight loss of appetite. Sub-acute infections can cause inflammation of the intestines and reddening of the fins. Large fish or brood stock may develop the typical symptoms of the disease in its acute form. There are swellings or 'furuncles' which can occur anywhere on the fish's body. The swellings, which contain a reddish pus, may burst either before or after death, releasing a mass of bacteria to the water and spreading infection.

Treatment If the fish continue to eat, the disease can be treated with an antibiotic mixed with the food. The chemical compound most commonly used has been oxolinic acid made up as a commercial preparation of 1g oxolinic acid per 5g of powder base. This given for 10 days at a daily rate of 50mg/kg of body weight of fish. The bacterium has developed an increasing degree of resistance to oxolinic acid and other methods of treating this disease are being used. These include a return to sulpha drugs as well as more general use of oxytetracycline.

Bacterial septicaemias   These diseases are caused by bacteria belonging to the aeromonas and pseudomonas groups. They are

167

universally present in most surface waters holding fish. They are unlikely to cause disease unless the fish are stressed. Their activity is similar to furunculosis in that disease symptoms seldom appear below 10–12°C, although fish can become infected at lower temperatures. Infections are usually by *Aeromonas liquifaciens* or *Pseudomonas fluorescens* which produce much the same symptoms in infected fish.

Symptoms Surface lesions on the body, sometimes produced by bursting furuncles but usually first observed as open sores.

The diseases respond to the same treatments used as a preventative and as a cure for furunculosis.

Vibriosis This disease is a haemorrhagic septicaemia caused by the bacterium *Vibrio anguillarum* (and possibly other *Vibrio* spp) which is present world-wide, generally in marine or estuarial environments. It can infect salmon in fresh and salt water but outbreaks appear to be more frequent and damaging in some sea areas than in others and is more prevalent in waters which reach comparatively high temperatures. Some strains are observedly more virulent than others. At least two strains have been recognised in Pacific salmon, one of which may infect the fish in fresh water but produces no pathogenic effects until they are transferred to the sea. The other infects the fish during their sea life.

Symptoms The fish cease to feed and become lethargic. Haemorrhagic areas appear in the skin and there is a reddening at the roots of the fins, the vent and sometimes in the mouth. Bleeding occurs in the gills and intestine. Deep red sores may appear on the body. Outbreaks of the disease in acute form, such as can occur among young pink or chum salmon reared in salt water in shore-based tanks, may produce no external symptoms other than large-scale mortalities.

Treatment Similar to furunculosis. A vaccine given by immersion has provided a measure of protection.

168

Bacterial kidney disease (BKD) This disease is caused by a *Corynebacterium* spp. It affects salmonids in Europe and North America in both fresh and salt water. It is a serious disease of Pacific salmon held in pens or cages in the sea. It can be carried over from fresh to salt water. Previous infection during parr life can cause large losses of smolt with damaged kidneys as they fail to stand up to the osmotic stress of transfer to the sea. In salt water the disease generally appears during the first winter.

Symptoms Chronic infections of young fish in fresh water may go unnoticed until casualties occur after the fish are transferred to the sea. Smolt which die will be found on dissection to have whitish lesions in the kidney and bleeding from the kidney and liver. Infected fish in sea cages may cease to feed and swim near the surface. They can appear dark in colour when viewed from above and show swellings on the sides. The eyes may bulge. There can often be no external symptoms and large fish in salt water may continue feeding actively, until they suddenly die for no apparent reason.

Treatment The disease cannot be treated in the sea, although some temporary arrest has been achieved by orally administered antibiotics.

Pancreas disease (PD) This disease made its appearance in the mid-1970s. The condition now known as PD causes degeneration of the exocrine pancreas which is the part of the pancreas that produces digestive enzymes. It is distinguished from the disease known as infectious pancreatic necrosis by the absence of any detectable causative virus and because it only appears in the sea. Outbreaks can occur at random on new or established salmon farms often in post-smolt in their first year in the sea but also in older fish.

The fish suddenly cease to feed and may collect in groups close to the surface. Some fish seem unable to maintain equilibrium. The more sickly fish quickly lose weight and become like kelts with darkened skin. The

169

only apparent tissue damage is in the pancreas. The weakened condition of the fish renders them susceptible to secondary infections by bacteria such as vibrios causing more general haemorrhagic septicaemia.

It has been suggested that the disease is a result of dietary deficiencies, possibly of vitamin E, but it can appear in different sea cages on the same site, where all the fish are given the same feed. The disease can also affect the fish in the same cage differently in separate outbreaks. The fish cannot be treated with antibiotics because they are not feeding.

The greater proportion of the affected fish will recover over a period which can be from weeks to several months. Salmon which have recovered seem to develop immunity and do not produce symptoms when further outbreaks occur in other age groups of fish on the same farm. Up to about 10% of the stock in a cage may have to be culled, but the loss of growth is of more economic concern to the fish farmer.

Like other seriously damaging diseases PD seems more likely to occur in some locations than in others. So far attempts to relate outbreaks to bad husbandry or to the environment have proved negative, but the failure to find any causative organism associated with damage to the exocrine pancreas does give credence to the suggestion that diet, related possibly to average water temperature in a given sea area, may be the initiating mechanism.

Hitra disease This disease takes its name from the Atlantic salmon farm in Norway where it made its appearance in the 1980s. The causative organism is thought to be a species of *Vibrio*. Haemorrhagic tissues, particularly in large fish, can show a complex bacterial mix. Like most haemorrhagic fish diseases, stress induced by intensive production probably plays a large part in spreading the symptoms of the disease. If husbandary and environment can be improved the fish respond to

170

treatment and can live with the disease. Bacteria are parasites and it is not in their interests to kill the host.

Myxobacterial diseases
Bacterial gill disease

The cause of this freshwater disease in salmon fry and parr is an infection of the gill filaments by a myxobacterium. It is usually accompanied by mucus clogging the gills which may be the result of irritation from finely divided solids in the water supply or from dust in starter or crumb feed.

Treatment Avoid over-crowding. Filter the water supply to remove irritant debris and shake out the dust from starter feed using a very fine mesh domestic sieve. Short-term baths in a solution of a bacteriocide such as chloramine T or benzalkonium chloride (BKC). The recommended concentrations vary in relation to pH and reference should be made to the appropriate manufacturer's instructions.

'Coldwater' disease and columnaris disease

These myxobacterial infections are common in hatcheries rearing Pacific salmon fry. They are caused by bacteria of the *Cytophaga* spp. and by *Chondrococcus columnaris* respectively.

Treatment Similar to that given for furunculosis. Short-term baths in a bacteriocidal solution.

Saltwater myxobacteriosis

Caused by marine myxobacteria of the *Sporocytophaga* spp. The disease occurs in Pacific salmon and is likely in Atlantic salmon farmed in Western USA and Canada. It can cause epizootic outbreaks among salmon farmed in sea cages or pens.

Treatment As for 'coldwater' disease.

Salmon sea louse (Lepeophtheirus)

This is a large copepod shaped like a miniature crab which lives on the skin of the host fish. The females are about 3–5mm in diameter and are distinguished by a pair of egg sacs. The eggs are probably shed in the summer or autumn. Free-swimming nauplius and metanauplius stages are followed by the first chalimus

171

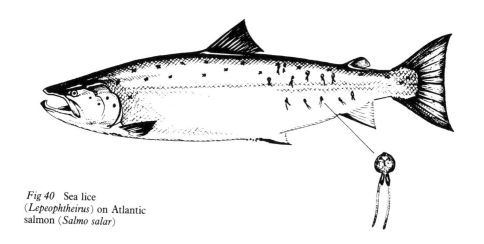

Fig 40 Sea lice
(*Lepeophtheirus*) on Atlantic
salmon (*Salmo salar*)

stage when the copepod attaches itself to a host fish.
Parasites in the chalimus stage can be seen on the fins
of the fish are and about 1·5 to 3mm in length. Several
moults occur during the chalimus stages and are
followed by numerous others before they attain their
maximum size. The parasites feed on particles of skin
and possibly blood drawn from the host fish.

Sea lice can probably migrate from one fish to
another and possibly to farm fish from wild fish in the
vicinity of cage sites near river estuaries or other places
where wild salmon congregate in coastal waters. The
larval stages appear to be able to migrate over
considerable distances carried by tidal currents.

Massive infestations can occur on farm salmon held
in shore tanks as well as sea pens or cages. The fish are
weakened and eventually die. The action of the sea lice
themselves on the skin; or the fish, frantic with
irritation, rubbing themselves against the nets or tanks,
causes lesions which become entry points for bacterial
infection. In some locations sea lice are the greatest
single cause of potential loss to salmon farmers.

Lepeoptheirus can probably complete its life cycle
inside a salmon cage, probably occupying the net mesh
during the larval stages and nets may have to be

172

changed every three weeks. Chemical treatment is difficult, particularly in large cages. Various methods have been tried. These include lifting the nets to reduce the enclosed volume of water and putting a polythene sheet completely under them or round the periphery to form a skirt. An added complication is that the fish are distressed by the medication and have to be prevented from injuring themselves or jumping out of the cage. Aeration may be needed while treatment is being given.

Treatment Organophosphorous compounds are the most effective chemicals tested so far. The commercial preparations are powders soluble in water from which a stock solution can be made up. The manufacturer's instructions give directions for dilution and strength. Internal medication which renders the tissues of the host unattractive to the parasite has been tried out and may prove successful.

Caligus This parasitic copepod is similar in many respects to *Lepeophtheirus* and can have equally serious implications for salmon farming in the sea. It differs from *Lepeophtheirus* in appearing to prefer offshore to coastal waters and is smaller as well as being more oceanic. There appears to be no build up of a local population on farm sites and infestation comes in waves from outside the cages. *Caligus* causes intense irritation and the fish frequently injure themselves by jumping or by abrasion from the net mesh.

Treatment Similar to that given for *Lepeophtheirus*.

Freshwater parasitic protozoa The young salmon grown-on in hatcheries or smolt farms during the freshwater period of life, prior to transfer to the sea, should be too valuable to be reared in conditions which can expose them to parasites. Bad hygiene and poor quality water can sometimes lay the fish open to attack. *Costia* is a microscopic pear-shaped protozoan living on the skin of fish. *Chilodonella* and *Trichodina* are larger more slow moving protozoa. All three can

173

sometimes be found on salmon parr kept in overcrowded conditions.

Treatment Baths in formalin or formalin and malachite green.

Fungal disease
Parasitic fungi

A common cause of loss in fresh water, particularly of eggs and alevin (yolk-sac fry) is infestation with *Saprolegnia* spp. the spores of which are present in the water.

Treatment Malachite green (zinc-free) has been generally used to control *Saprolegnia*; either as a dip at 1 : 15,000 (67ppm) with fish immersed for thirty seconds or as a bath (1 : 500,000 = 2ppm) for fry and parr. The bath concentration can also be used as treatment for incubating eggs after they have reached the eyed stage.

Ichthyophonus

This fungus occurs as a parasite in marine fish but can also invade the organs of farm salmonids, fed on infected sea fish, at both salt and freshwater sites.

Diseases caused by algae and dinoflagellates

Unlike wild fish, fish in enclosures or cages in the sea cannot escape from a sudden wave of poisonous or damaging material entering their confined environment. Dinoflagellates have caused massive losses of salmonids in sea cages. The fish may be killed by a nerve toxin produced by some dinoflagellates such as those forming the so-called 'red tides'. Blue-green algae have also been reported as causing losses.

'Fish kills' in cages or enclosures are not only caused by direct poisoning but may also be due to suffocation following the removal of all oxygen in the water by the respiration of algae during the night. The physical clogging of the gills of the fish can also cause suffocation, either by the algae or by mucus produced in the gills of the fish as a result of irritation.

Algal blooms can occur at any time in bright spring or summer weather and little is known of the causes. It

174

seems likely however that sudden changes in water temperature or salinity may act as triggers. Nothing can be done to protect the fish in cages or enclosures sited in an area subject to poisonous algal blooms. The only course of action is to move the cages to a safer place.

Immunization of farm fish

Methods for the immunization of salmon against specific diseases so far tried out have been intra-peritoneal injection and hyperosmotic infiltration. The diseases involved are furunculosis and vibriosis respectively. Both methods yield positive results but the difficulty of inoculating large numbers of fish could make this method of immunization a doubtful economic proposition for commercial salmon farmers. Hyperosmotic infiltration when used to immunize coho and chinook salmon fry has given over 80% survival compared to less than 30% in untreated fish after transfer to the sea. Commercial vaccines are available for both immersion and injection.

Hygiene and disinfection

Disinfection of live fish in fresh water

It should not be necessary to treat fish for external parasites, if the water supply is free from any organic pollution and saturated with oxygen, provided the fish are not overcrowded and a good standard of hygiene is maintained on the unit.

The condition of the fish and their environment must be taken into consideration, particularly the chemistry of the water. If the oxygen content is low, aeration will be essential during treatment. The chemical used in dips or baths can be more or less active according to the pH, and in acid waters may reach toxic levels even when made up at recommended concentrations. The stress of treatment may itself cause an unjustifiable increase in the casualty rate.

Advice should always be sought if a fish farmer is not completely certain of the diagnosis of the disease and the correct treatment. It is essential to make sure

175

that the fish are not suffering from more than one infestation of parasites at the same time. Gill parasites should always be treated first. As a general rule, the fish should be starved for one or two days prior to treatment as this will reduce the ammonia content of the water.

The strength of the solutions used to make up disinfectant dips and baths should be carefully checked. It is usually too late to do anything about it when the fish begin to show signs of distress.

Dips The fish are held in a net or sieve and lowered into a concentrated solution, usually for not more than about thirty seconds.

Baths The fish are immersed in the chemical solution for up to one hour (sometimes longer). The oxygen level must be monitored in static water and aeration should be available for use if needed.

When the fish are in tanks or raceways a calculated 'drip' of the chemical at the correct concentration can be added to the water supply. This is expensive but treatment can usually be given in fresh water, without the stress of capturing the fish. Low-level disinfection can be provided over a longer period. Care must be taken with the effluent water containing the chemical solution as this can be a source of pollution.

Formalin A most useful compound which is commonly used for the chemical treatment of external parasites on fish and is a 40% solution of pure formaldehyde, uncontaminated by paraldehyde which is poisonous to fish.

Formalin should be carefully handled as it is a respiratory irritant. It must be completely mixed so that the concentration is evenly distributed through the fish tank. In order to make sure that this happens, a little malachite green can be mixed with the formalin as a tracer.

The concentration in the fish tank should not be greater than 1 : 5,000 and the concentration should be

176

reduced to 1 : 6,000 if the water temperature is over 15°C. The time of treatment should be not more than one hour.

Formalin reduces the oxygen in the water and aeration should be provided during treatment.

Formalin and malachite green Stock solutions for control of external protozoan parasites on salmonids.

Water above 15°C	*Water below 15°C*
1 part malachite green to 300 parts formalin	1 part malachite green to 200 parts formalin
Mix together—1 litre of formalin (37% formaldehyde) with 3·3g malachite green	Mix together—1 litre of formalin (37% formaldehyde) with 5·0g malachite green
Dilution for treatment 0·015ml/litre 15ml/m^3	Dilution for treatment 0·020ml/litre 20ml/m^3
Final dilution Formalin 15ppm Malachite green 0·05ppm	Final dilution Formalin 20ppm Malachite green 0·1ppm

Treatments with these concentrations of active ingredients should not exceed six hours in tanks with running water, or with aeration. Look out for signs of stress in badly parasitized fish.

Hygiene Like any other domestic animal, fish do best if kept under hygienic conditions. This applies particularly to salmonids which are cold water fish and will not tolerate an adverse environment.

Hatcheries Suspended solids should be filtered out of the water supply. Debris should not be allowed to accumulate at the bottom of troughs below egg trays. It is particularly important to prevent any accumulation of waste food or faeces in troughs or small tanks when these are used for initial fry feeding. Hatchery equipment should be disinfected and stored dry when not in use.

Fry tanks The bottom of tanks and outlet screens should be kept thoroughly clean. After a season's use tanks and screens should be scrubbed out with a disinfectant solution and left dry. The insides of fry tanks should be given a coating of a neutral anti-fouling paint during the off-season to prevent algal growth.

General equipment Separate cleaning utensils should be provided for each tank and kept in a weak solution of disinfectant. Buckets, transportation tanks and any other equipment which comes in contact with the fish should be disinfected after use.

Food hygiene A common cause of loss is through bad food being given to the fish which has either deteriorated during storage or been prepared under foul conditions with dirty equipment. It is particularly important to store fish food with as good or better attention to hygiene than is paid to human food. This applies to both fresh, wet or deep-frozen foods and to dry foods.

Disinfectants used on fish farms will cause pollution if they are released undiluted into water courses.

Potassium permanganate. A fairly concentrated solution 1 : 50 is a useful, non-corrosive disinfectant for troughs, tanks and equipment.

Saltwater shore-tanks The general degree of hygiene that can be applied in shore-based tanks, using pumped sea water, is the same as for freshwater tanks. The accumulation of marine growths creates special problems as these not only impede the flow of water but also collect waste food and faeces. There is no way round this problem other than to design the unit as simply as possible so that it can be kept clean and to use such non-toxic anti-foulants as may be available.

Sea cages and enclosures There is little that can be done to keep enclosures from becoming fouled with accumulated waste food and faeces. These may eventually have to be dredged or pumped out. Cage nets have to be changed periodically depending on the marine growth in the area. Marine plants will grow on the nets in spite of anti-fouling treatment. They are best cleaned off by immersing the nets in fresh water and then if possible getting them thoroughly dried out.

Cage nets on smaller cages can be made with a 'cod end' or tapered bottom ending in a funnel of netting. This is closed with two or three turns of cord and a pin. The end of the funnel is connected to the cage walkway by a rope. Debris collects in the funnel at the bottom of the cage and from time to time it can be hauled up and emptied.

Dead fish have to be removed and a count kept of casualties. In large sea cages this may only be possible by diving. All large salmon farms using sea pens or cages should have at least two members of staff trained as scuba divers.

Fish health Salmon farmers should understand that they are dealing with wild fish about which relatively little is known. This applies particularly to the behaviour of the different species during their sea-life and the marine environment essential to their proper growth and well-being. Coho and chinook are more tender than Atlantic salmon and easily lose their scales when handled or from abrasion on the sides of pens or cages. This can render them prone to infection but as with all salmonids being farmed commercially in the sea, including the hardier Atlantic salmon, the main danger is from specific diseases to which particular species are prone when overcrowded in an alien environment.

The geographical location of a sea farm is certainly of importance to the health of both Atlantic and Pacific salmon. Some sea areas may provide the conditions in which disease bacteria or crustacean and protozoan parasites can thrive. The average water temperature can be too high or too low. The summer day-length can be too short. Atlantic salmon certainly spend most of their marine life in waters north of the Arctic circle. The races in the most northerly rivers spend their

179

whole life cycle in both fresh and salt water where the nights are six months long and there are 24 hours of daylight for most of the summer.

Salmon are being farmed commercially in the coastal waters of temperate seas far from their marine habitat and there is little that can be done to compensate for the loss of free, oceanic life. It is evident, however, that the risk of disease depends to a great extent on the local environment in which the fish are grown. A great deal has yet to be learned about animal welfare in salmon farming and the knowledge applied in practice. Prevention is better than cure and salmon farmers must sooner or later learn to live with and control their fish diseases by good husbandry.

Useful chemicals for controlling salmon diseases

Chemical	Purpose	Method
Benzalkonium chloride (BKC)	Disinfectant	Bath or dip Use depends on pH
Chloramine T (sulphonic acid-free)	Myxobacteria. Gill disease External protozoan parasites	Bath or dip Use depends on pH
Copper sulphate + acetic acid 80%	Bacterial gill disease Fin rot	Bath or dip
Di-n-butyl tin (butyl tin oxide)	Flukes and worms Internal parasites	25mg/kg fish weight for 5 days
Formaldehyde 40% (Formalin)	External protozoan parasites	Bath or dip
Formalin/malachite green	as Formalin	Bath or dip
Iodophors	Disinfection Treatment of eggs	Bath or dip Dependent iodine content
Malachite green	Fungus. Saprolegnia Eggs and fish	Bath or dip
Nifurpirinol. Nitrofuran (Furanace. Nifurprazine)	External protozoan parasites Haemorrhagic septicaemias	Bath or dip Oral with feed
Organophosphorous Nuvan. Neguvon	'Sea lice' Lepeophtheirus Caligus	Immersion
Oxolinic acid (Aqualinic powder)	Furunculosis Haemorrhagic septicaemias	50mg/kg fish weight for 10 days
Oxytetracycline	as for Oxolinic acid	50mg/kg fish weight for 10 days
Sulphamerazine	as for Oxolinic acid	200mg/kg fish weight for 14 days

Note: Treatment with antibiotics in fish feed must cease 3 weeks before slaughter for market.

13 'One-shot' salmon production

This system of Atlantic salmon farming is so-called because it involves stocking the cages with very large smolt and harvesting the fish later in the same year, without grading. The system is based on the use of the largest 'high-seas' type cages which should be anchored in sea areas where the water warms up quickly in the spring and remains above 10°C for most of the winter.

The cages must be stocked with very big smolt reared in heated water, or the largest obtainable S2 smolt from freshwater cages in comparatively warm lakes where the fish have continued to grow through the winter. The average weight of the smolt should be at least 150g. Fish from geothermally warmed hatcheries weighing 250g or more are excellent. The cages are stocked in April/May depending on the sea temperature and the availability of acclimated smolt.

The growth of the fish is monitored by sampling at regular intervals. When a pre-determined average weight (2–2·5kg) has been reached, all the fish are harvested and market-graded for size before being boxed and iced for fresh sale or quick-frozen for cold storage.

'High-seas' cage anchorages   The nets used on these cages are 10m (33ft) deep and they should be anchored in at least 15m (50ft) of water

at low tide. The cages will stand up to gale force winds and the wave action generated over a long fetch, but rough seas and strong winds can make servicing difficult and sometimes hazardous. On the best marine sites for this type of cage, exposure only prevents the fish being fed under exceptional weather conditions which occur on very few occasions in the year.

Husbandry Servicing large cages on exposed, off-shore anchorages, may require special equipment. A powerful workboat is essential. This can either be big enough to carry fish food and fish to and from the cages and to provide space for working gear on deck, or able to tow service rafts or barges and hold them in position alongside the cages.

During the final months of growth each cage of fish can take an average of $1-1\cdot5$ tonnes of feed a day (more if the cages are densely stocked), which has to be fed to the fish over a wide area. Various methods are used, including hand-feeding and portable

Fig 41 Water 'cannon' fish feeder: (A) feed hopper; (B) 'cannon' barrel; (C) flexible pipe to water pump; (D) swivel

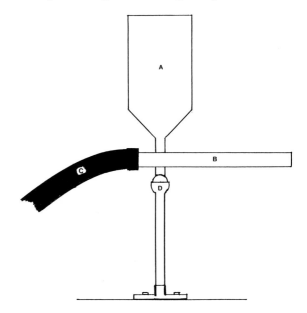

compressed-air gun feeders, but the best method so far developed is a water cannon.

A small high-pressure petrol-driven impeller pump is needed with a 2–3in. discharge port and a total head (distance over which a jet of water can be discharged) of 24–29m (80–100ft approx). The Yamaha Motor Co. of Japan make light, economical pumps of this type. The pump is bolted to the deck of a workboat or service raft. The weighted end of the intake pipe is hung overside in the sea and a flexible 2 or 3in. hose connects the pump to the 'cannon'. This has a metal or rigid plastic barrel, mounted on a ball-socket swivel which is fixed to a vertical stand mounted on the gunwale or deck of the boat or the deck of the service raft. A hopper, taking one 25kg bag of feed, funnels the feed into the top of the 'cannon' barrel. Feed is drawn from the hopper by venturi action and fires out of the barrel in the jet of water. The swivel allows the jet carrying the feed to be directed in a wide arc over the cage. If the hopper is loaded by hand with 2 loaders, the 'cannon' will deliver about 1 tonne in 50 mins from 25kg bags. The water 'cannon' will also feed moist pellets but these must be mechanically fed into the barrel.

Net changing

This can be a laborious operation when working with very large cage nets. It can be greatly assisted if a fishing boat with net hauling winches is available. Service barges or pontoons moored alongside are essential to provide a stable working platform. Knotless, polyethylene nets are available in 12mm and 22mm mesh incorporating strands of fine copper wire in the weave. The copper inhibits marine fouling to some extent and can reduce net changes to one per growing season when the mesh size is increased.

Treatment for disease or parasites (sea lice— Lepeophtheirus)

This can be administered in the cage by raising the net to concentrate the fish in shallower water which is then surrounded by a curtain of plastic sheeting. An

183

alternative method is to transfer the fish into a smaller, service cage, in a collar of pontoons, moored alongside. The service cage is enclosed in plastic sheeting. An air pump can be used to aerate the water during treatment.

Transfer The removal of live fish from the main cage net, either for treatment or pre-market selective grading before slaughter, involves concentrating the fish into a space from which they can be caught-up or brailed. The cage net is raised to reduce the depth, one side can then be detached from the flotation collar and brought across the cage. In very large cages, the top of the net will have to be carried over on a dinghy which can be swung into the cage, or linked to floats before it is hauled across. An alternative method is to use a seine net, hauled inside the raised cage, to enclose the fish. Plastic tubing or channelling can be used to transfer the fish into a service cage.

Harvesting and slaughter The fish brailed from the main cage are selected for market size/weight and transferred into tanks on a service barge where they can be killed by CO_2 (carbon

Fig 42 Collecting salmon from a large sea cage

184

dioxide) released into the water from cylinders. Undersized fish are transferred to a service cage and from there back into the main cage.

Cage servicing and inspection   It is essential to have at least two sea-farm staff trained as divers. The value of the fish in large cages makes it imperative to inspect the nets at frequent intervals. Also the divers must swim through the fish in the deep cages to check their condition and to remove any casualties for examination and recording.

Model 'high-seas' salmon farm (see *Fig 43*)   The farm shown uses two hexagonal 'high-seas' cages (AA) each enclosing 6,650m^3 of water. The cages have a combined potential production of 70–100 tonnes in a growing season. A service platform or raft (D) is anchored close to but separate from the cages. Service cages (BB) 6m × 6m are moored on each side of the barge. Two mobile working pontoons (CC) 12m in length and 2m wide are moored to the service cages or can be anchored separately. The mobile pontoons can be moved and moored alongside any straight section of the hexagonal cages to act as working walkways.

The service platform has water 'cannons' (SS), with food hoppers and pumps, mounted at each end. The service raft provides a stable base from which to feed and tend the fish. All gear, feed and medication is

Fig 43 'Super' cage site plan: AA flexible 'high-seas' cages, 16 m sides; BB service cages, 6 m × 6 m; CC mobile pontoons; D floating platform; SS water feed 'cannons'

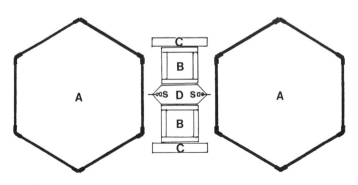

185

brought out from the shore in a workboat. The small cages are both separately anchored as well as being moored to the service platform. The moorings can be cast-off and the anchor cables are long enough to allow the small cages to be brought alongside the main fish cages.

Stocking system for Atlantic salmon

One cage can be stocked in April with 25,000 very large smolt reared in heated water. Harvesting this cage can start in November and continue through to February at an average rate of about 4 tonnes per week. The second cage is stocked in May with large S2 smolt. Harvesting can start in March of the following year and continue into the summer. This system provides fish for market and a continuing cash flow over a long period. The fish are harvested simply on a basis of size/weight. No attention is paid to whether they are grilse or pre-salmon.

14 Salmon ranching

Ranching of salmon starts by releasing the young fish at or near the stage they would naturally migrate to sea. The fish then complete the sea-feeding part of their life cycle on free-range and return when mature to the river where they were released. In order for ranching to succeed, enough salmon must survive to return to the point of release, at a place where they can be captured in commercially profitable numbers. The return of the salmon not only depends upon natural survival, but on their not being caught by other fishermen, either on the high seas or on their way home.

Ranched fish have complete freedom to migrate to and from their marine feeding grounds. The improvement over nature is obtained by artificially incubating and hatching the eggs of the fish, and rearing the young through the period of greatest natural loss in the wild. It must not be confused with what is called 'enhancement'. This means planting the eggs, fry or parr of salmon in sections of the river which the adult fish cannot reach on their spawning migration, or which are unsuitable as spawning ground but can provide useful rearing areas for young fish.

The development of artificial spawning channels for

187

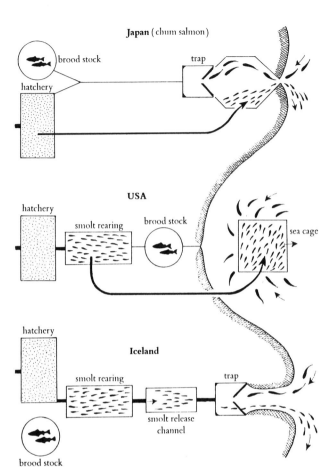

Fig 44 Salmon ranching methods

some species of Pacific salmon is a half-way stage between full ranching and enhancement. A typical spawning channel is an artificial canal made beside the natural river, or across the loop of a wide bend. The bed of the channel is covered by graded gravel of the correct size to be used for redd or nest construction by the selected species of salmon. The inflow and outflow of water is controlled by sluices or valves, and the

188

length of the channel may be divided up by weirs and screens.

A number of adult fish returning on spawning migration is counted into the channel, where they spawn naturally. Alternatively the fish can be stripped and the fertilized eggs planted in artificial redds or nests made in the gravel. The use of spawning channels has substantially increased the marine harvest of some species of Pacific salmon. The method is only of real value for the species which migrate directly to sea as fry (pink and chum), or leave the spawning area in a river to feed in lakes (sockeye). If the young fish have to be fed artificially, it is better and more easily done in tanks.

Marine growth   Feeding on the 'sea-range' depends on conditions in the environment, as it once did for cattle and sheep on the open range. Salmon production in the sea is dependent on the changes in the ocean currents and temperature gradients that make up the shifting climate of the sea. All ocean-going salmon follow the animals on which they feed, whether their food consists mainly of crustacea in the plankton or species of shoaling fish. Salmon bred in rivers at the southern end of their range go north to feed in the sea, and may have a long journey to and from their marine feeding grounds, compared to members of the same species from northern rivers. This is an important consideration in ranching because the fish will be in better condition if they have a short return migration. They will burn-off less fat under conditions of semi-starvation while they are travelling back to their parent rivers.

Pacific salmon   A good deal is known about the marine environment of the Pacific salmons, because the most valuable species are fished for commercially in their feeding areas. The sea-feeding areas of Atlantic salmon are less well defined. Drift netting in the North Atlantic, outside

189

Table 8 Pacific salmon

Species	Max. wt. kg	Av. wt. kg	Av. age yrs	Parr life	Marine food	% Commercial value
Pink O. gorbuscha	5·4	2·7	2	Nil	Plankton, small fish	40
Chum O. keta	15·9	5·9	3–6	Nil	Plankton, small fish	32
Sockeye O. nerka	6·8	3·6	4–6	1–2 yrs (in lakes)	Plankton, small fish	100
Coho O. kisutch	11·3	4·0	3–4	1–2 yrs (in rivers)	Mostly fish	88
Chinook (Spring) O. tshawytscha	45·0	8·0	3–8	Some months (in rivers)	Mostly fish	72

The older average ages are for fish belonging to colder rivers in the northern part of the range of the species. Sockeye are the most valuable species and their commercial value is taken as 100%.

territorial waters, has been banned by international agreement between the main fishing nations. Only inshore feeding grounds, such as those off the coast of Greenland and north Norway, are well documented from the results of commercial fishing.

The richest feeding grounds in the northern seas are in areas where the cold, less saline water from the melting ice-cap meets the warmer, strongly saline water brought up by the ocean currents circulating from south to north and west to east. Changes in the pattern of the meeting of these waters, such as cooling due to an increase in the influence of the Arctic ice-water and a weakening of the warm currents, could greatly influence the quantity of the food animals available to sea-going salmonids. The marine food supply is, of course, not only of vital importance to ranched salmon, but to all stocks, including those which are completely wild and not managed in any way by human beings.

Breeding and rearing salmon for ranching

Ranched species need not only be those which have a parr-life period in fresh water. Very large numbers of chum salmon fry are hatched artificially and released in Japanese waters to return as adult fish. Atlantic

190

salmon, as well as the Pacific species, can be used for ranching; but rearing smolt intended for ranching requires the employment of rather different techniques to those used on fish intended for farming. The smolt to be released on free range must be able to survive to adult life in worthwhile numbers in the wild, without artificial feeding or protection from predators. The larger they are when they go to sea the better the chance they have of surviving.

The feed given to the fish should be similar in protein and fat content to the natural diet and the use of sources of protein containing carbohydrates should be avoided. Differences in body chemistry between artificially reared and wild smolt, or pre-smolt parr, have to be rectified after release to adapt to the change of a natural diet. It has been shown that the body chemistry of young, hatchery-reared coho takes about three months of wild feeding to revert to its natural state.

It is evident from the results of Atlantic salmon ranching in Iceland that large smolt survive better than small smolt. This applies not only to older smolt but to smolt in the same year class. This indicates that there is an advantage to be gained in using S2 Atlantic salmon smolt for ranching.

The main causes of loss in the wild are starvation due to competition for food and living space, and predation. The disadvantage in artificial rearing is that it takes away the important element of natural selection which weeds out the weaklings and leaves the strongest and most vigorous fish to survive and become wild smolt.

Releasing pre-smolt Atlantic salmon parr and smolt

Pre-smolt reared in freshwater cages in lakes should be transferred to a special 'release' cage which is then towed down to the outlet stream from the lake. Feeding should stop at the first sign of silvering. A release cage should be comparatively long and narrow with the deepest part of the net at one end. The

191

shallow end is taken into the mouth of the stream and opened to allow the fish to migrate freely downstream when they are ready. Large S2 smolt should be used and if they are to be marked this should be done with microtags in the previous autumn or fall.

The system developed in Iceland for releasing Atlantic salmon smolt has proved successful. The fish are removed from their rearing tanks to release ponds shortly before smoltification is complete. The release ponds are constructed by excavating an area beside the lower part of the river to which the adult fish are expected to return. The ponds are fed with a freshwater supply at the upstream end. The downstream end is open to the sea or a tidal estuary and is usually controlled by a sluice-gate. Salt water enters the pond at high tide. Smolt can shift gradually into water of increasing salinity by moving down the pond and are free to go to sea when they feel inclined. The same pond, with a trap at the upper end and a diversion fence across the river, can be used to retain and capture returning adults.

Most Icelandic ranching is done with S1 smolts, as all the rearing stations accelerate parr growth with warmed water. Smolts intensively reared in water at $12-13^{\circ}$C must spend at least a month in water similar to the expected sea temperature at migration before being transferred to the release ponds. Fish reared indoors in artificial light and heated water need a longer period to adjust to natural day lengths and water temperatures.

Survival of ranched Atlantic salmon in the sea

It has not yet proved possible to arrive at a completely accurate estimate of the survival of ranched Atlantic salmon from smolt to returning adult, except possibly in the Icelandic fisheries. The Icelandic salmon ranches use a very small wire micro-tag which is inserted by a special gun into the nose of the parr. This type of tag does not inconvenience the fish in any way. It is quite

192

invisible on the body surface and leaves no scar. All returning fish have to be checked for tags with a sensitive metal detector or by X-ray.

There is no off-shore netting for salmon in Icelandic waters, so the catch or count of returning adults made in rivers can give a true indication of the percentage survival of the fish from smolt to adult. Large Atlantic salmon smolts give an adult return of 15%.

Returns of tagged Baltic salmon smolt released from Swedish rivers have averaged 12%. In one experiment the return was as high as 40%. The return of artificially reared smolt in Scotland and Ireland has been about 6% but the results are not truly relevant because of the large proportion of adult recaptures made by drift nets in coastal waters that go unrecorded. The true survival percentage of artificially reared smolt from rivers in these countries is probably not much different from that demonstrated in Iceland.

Ranching has been tried out with varying success using the species of Pacific salmon which spend some time as parr in fresh water before migrating to the sea. The difficulty is that these species can be fairly well advanced towards spawning by the time they return to their parent rivers. Releases have been made from cages or enclosures in the sea in the expectation that sufficient adults could be recaptured in the vicinity of the cage areas. The most successful species for ranching seems likely to be coho which has a life cycle not unlike that of Atlantic salmon.

Rivers suitable for ranching   It is a waste of time and money to try to start ranching in an area where there is already large-scale legal or illegal salmon fishing with nets in the sea, either drift nets off shore, fixed gill nets or trap nets. The salmon must have a clear run home from their sea-feeding grounds, without being exploited to any extent by catch fishermen. There are not many places where ranching is possible, particularly if private fishing

193

rights are neither respected or protected, and governments are not prepared to enforce national or international conservation laws.

If the fish can get a safe passage home it is not necessary to ranch on a large river. It is quite feasible to induce a return of adult salmon to a small stream (or possibly to a place in the sea), provided the fish will return to a circumscribed area where they can be easily captured. Small streams that are fed by a lake can be made more attractive to returning salmon, and the entry of the fish can be facilitated, by artificial spates created through a sluice-gate in a low dam.

Homing Ranching is based on the ability of salmon to return to their parent rivers on spawning migration. It has been demonstrated beyond doubt that a sense of smell is the means by which salmon recognize the waters in which they were hatched and from which they migrate to sea. The imprint of the home-base must be developed by some species in the very short time they spend as fry in fresh water. Evidence from the results of transporting smolt to alien rivers before they are released, indicates that the imprinting process may take longer and that mistakes can subsequently be made by the returning adults.

The olfactory basis for homing may create problems for ranching, if smolt are not released in the river in which they were reared. It would be possible for fish to make mistakes if the volume of water leaving the release stream closely resembled the water coming from another neighbouring catchment area. There is also the possibility of the olfactory stimulus from the home water being masked by some source of pollution.

Homing to a particular part of the coast must be governed by other means than the sense of smell. If smolt, released in the sea, are expected to return to the release area as adults, it seems likely that something more positive than the general marine guidance system may be needed to get them back to the right place.

194

Trapping and counting   A simple inscale trap is needed to take returning adults. The returning fish have to be guided to the trap entrance by grids forming a fence across the river. More than one trap may be needed over the width of larger rivers. Fish can be directly counted in the trap. The only satisfactory counters for salmon are those of the so-called resistance type. These are based on the principle that the body of the fish contains a hypertonic salt solution and has less resistance to the passage of electricity than a similar volume of fresh

Fig 45 Early days of salmon ranching in Ireland: fishing weir with traps

Fig 46 Collecting salmon from a fish trap

195

water. The fish are guided to swim through a tube or an open channel which has metal electrodes buried in the sides and base. A very small electric current passes through these electrodes and this increases when a fish goes through the counter. The increased current is amplified and is used to operate a recorder.

The future of salmon ranching

Ranching salmon, which is a means of taking advantage of the natural growth of the fish in the open sea, is an attractive proposition, and one of the most exciting possibilities open to salmon farmers. The future success of any kind of ranching, and the status of the world's stocks of sea-going salmonids, depends upon the perhaps unattainable goal of mankind; co-operation between nations in the conservation and utilization of the world's natural resources.

15 Prospects for salmon farmers

Salmon farming in common with most forms of
aquaculture is a high-risk business. The risk is
compensated by the high return on investment.
Success or failure is rooted in planning. A common
cause of disaster is to settle on a bad site and follow
this with over-capitalization. Any project should first
be examined from the point of view of marketing. The
basic questions are of demand, distance to market, cost
of marketing and sale price.

When starting to plan a project, it is vital to obtain
expert advice from people who have had personal,
practical experience of salmon farming. This applies to
accountants as well as technical consultants. Sites have
to be found and surveyed. A renewable lease has to be
agreed, or some other arrangement providing security
for exclusive tenure of both sea and shore bases. A
financial assessment can then be made, based on
technical requirements in terms of capital investment
and fixed and variable costs over the years leading to
full production.

Work forces   Sea farming is not labour intensive. Production per
employee year is a measure of efficient management
and should be about 25 tonnes on large salmon-

197

fattening farms. A hatchery and smolt rearing unit can be easily operated by two workers.

The capital invested in order to produce a given tonnage of fish per year can vary, according to design and circumstances, from less than half the gross value of each year's fish production, at the most profitable end of the scale, to about four times the value of a year's fish production at the least profitable end. The most expensive production system, in terms of return on capital, is a shore-based salmon farm, with a pumped seawater supply. The lowest cost units are medium-sized sea cages moored in separate flotillas, on a sheltered site.

The capital cost per tonne of annual production varies according to the design of the unit. There are no basic criteria for profitability. Large units can be over-complicated in an attempt to achieve sophistication and consequently over-capitalized. A common failing is that actual production never reaches the design capacity of the unit and the farm is always under-producing. The capital cost of a sea farm, producing a given tonnage of salmon in salt water, can be as much as five times greater than for another farm producing the same tonnage, but the cost is to some extent a reflection of the location and it may not be possible to construct a low-cost unit on a particular site.

An analysis of the figures from working farms indicates that there is a relationship between the gross value of an average year's production of fish and the current capital value of the unit. If the gross value of the product is less than the capital value then the farm may be losing money. The more profitable the farm, the greater will be the proportion by which the gross annual value of fish sales exceeds the capital value.

A short-fall in actual production compared to planned production may be due to mistakes in the original design of the farm. It can also stem from a number of other failings, some of which can be corrected, while others are inherent in local conditions. If fish fail to make the expected growth, or take too long to reach market size, it may be due to shortcomings in the site, but is more frequently the result of incorrect feeding or bad husbandry. Little or nothing can be done to improve the situation if it has come about as the result of the failure to make an adequate initial survey, and to properly assess the potential of the site.

Running costs The principle item in the cost of production of salmon is fish food. Dry-feed conversion rates should be between $1:1 \cdot 2$ and $1:1 \cdot 6$, depending on the site of the sea farm and the methods employed. Many salmon farmers use wet feed in the form of moist pellets for at least part of the year. Better and more even growth is claimed, particularly in winter. The water in the pellets is said to prevent osmotic stress in fish. Salmon in the wild obtain low-salinity liquids from the flesh of their prey. Dry feed can be transported and stored on site and the conversion rate is a good deal better than for moist pellets, but it can be assumed that a farmer intending to use wet feed has sited his farm close to where he can get fresh supplies. In isolated areas the lower conversion rate will be compensated by not having to import dry food over long distances. The use of moist pellets, made on the farm, is certainly worthwhile on sea farms close to the ports of landing for industrial fish.

The division of running costs is very much a matter of the location of the farm. An apportionment of production costs in saltwater farming of salmon in Europe is approximately as follows:

— Interest and depreciation   26%
— Fish feed   34%
— All other overheads   40%

199

Fish can be lost due to a wide variety of causes. They can be taken by fish-eating birds and large marine mammals, particularly seals. Cages can be damaged and fish escape due to the presence in the area of whales and basking sharks, or the activities of the crews on fishing boats.

The water may be polluted by the discharge of oil, chemicals and waste products from ships at sea or the effluent from shore-drains. It can become turbid or full of debris during storms or poisonous algal blooms can occur in the site area.

Failure of a power supply and mechanical breakdown of the pumps or other machinery can result in the death of fish through lack of oxygen. Structural failure due to faulty construction or lack of maintenance can result in the escape or death of fish. The following conditions can cause catastrophic losses of fish:

— storm damage;
— collision;
— sudden changes in temperature and salinity;
— algal blooms;
— turbidity;
— failure of the water supply;
— disease;
— and a whole variety of other unexpected and fortuitous incidents.

The premiums charged for fish farms are high because of the risks involved. Various ways are offered by which premiums can be reduced such as the 'franchise' principle in which the loss of fish has to reach a certain percentage of the total of the stock held on the farm at the time of the incident before the loss can be indemnified. Another method is the so-called 'deductible' which simply means that the farmer has to bear an agreed initial percentage of any claim.

For practical purposes fish farm insurance is best treated as a form of a disaster cover. The salmon farmer should decide the most likely causes of a possible catastrophic loss of fish on his farm, at a stage when it is most financially vulnerable. The farmer should then try to negotiate specific cover at an agreed rate for these eventualities only.

Cold storage and processing

Marketing freshly killed salmon in perfect condition is difficult and expensive. The fish have to be properly packed in ice, which means having an ice-maker on the premises. Melting ice only staves off deterioration for a relatively short time. Fresh fish are highly perishable and delays in public transport can result in a total loss. Insurance rates are understandably high.

Cold storage makes sales independent of immediate market considerations. The fish can be slaughtered, and processed when labour is available. Prices can be negotiated and agreed in advance with customers, and the fish sold direct from cold store. Bulk transport of frozen fish is cheap and free from the risk of spoilage.

The advantage of processing is that the value of the fish produced can be enhanced before they leave the farm. Processed fish does not have to pass through a chain of wholesalers, and markets can be found where there is a direct sale to retail outlets. A profitable

Fig 47 Arranging salmon to cool on the fish house floor

treatment for salmon is cold-smoking. There is a loss of about 25% in the gutted weight of the fish, but the value per kg is increased by more than 100%. The best quality smoked salmon is achieved by the method known as wet-brining in which the fish are immersed in a brine bath before being smoked, instead of the usual dry-salt treatment. Other processing includes cutting, and packaging in fillets and steaks.

The developing industry Back in the 1940s chicken was for special occasions. It was too expensive for everyday consumption. Then intensive production got under way and broiler birds appeared on the scene. Soon everyone with a little capital and a taste for the country life took up farming broiler chickens. Prices began to come down and economies of scale took effect. Now only the large companies are left, either marketing their own product or supplying young birds to small units where they are grown on and marketed under the direction of the parent company.

Farm salmon are inevitably following the same course as broiler chickens. Some small, family-run sea farms may continue to operate profitably if they can get supplies of fresh industrial fish to make up their own moist pellets and can afford the price of smolt, but the price of dry feed and problems of marketing could eventually concentrate the industry in the hands of the big producers. The Norwegian Government tried to hold down the size of salmon farms and keep them in the hands of local fishermen-farmers, but ways have been found round this well-intentioned restriction. One means for the small producers to survive is to farm under the direction of large companies which supply the smolt and collect the fattened fish for central marketing. The alternative is to form co-operatives. Salmon farmers, like fishermen, are individualists and may not take kindly to working as part of an organization governed by their erstwhile rivals, where all their affairs are open to scrutiny, but it can prove to be the only way to compete on level terms with the larger operators. The co-operative would be responsible for growing smolt on a central freshwater unit, for the collective purchasing of equipment and fish feed, and for marketing. It would be able to afford to employ trained staff and draw on expert technical advice.

It would be a great loss to see the small-scale fisherman or crofter salmon-farmer fail, as the industry at that level can maintain economic life in isolated areas which could otherwise become depopulated.

The final product The age of salmon is of no concern to the consumer. All that matters is the size, taste and colour that the housewife decides she prefers. She is the eventual arbiter of full development in the salmon farming industry.

Conversions to and from metric

1 in	= 25·4 mm
1 mm	= 0·0394 in
1 ft	= 0·304 m
1 m	= 3·281 ft
1 yd	= 0·914 m
1 m	= 1·094 yds
1 mile	= 1·609 km
1 km	= 0·621 mile
1 ft^2	= 0·0929 m^2
1 m^2	= 10·764 ft^2
1 yd^2	= 0·836 m^2
1 m^2	= 1·196 yd^2
1 acre (4840 yd^2)	= 0·405 hectare
1 hectare (10,000 m^2)	= 2·471 acres
1 ft^3	= 0·0283 m^3
1 m^3	= 35·315 ft^3
1 yd^3	= 0·765 m^3
1 m^3	= 1·308 yd^3
1 UK gallon	= 4·546 litres
1 US/Canada gallon	= 3·785 litres
1 lb	= 0·454 kg
1 kg	= 2·205 lbs
1 UK ton	= 1·016 tonnes (1016 kg)
1 US/Canada ton	= 0·907 tonnes (907 kg)

Some useful equivalents

Unit	Equivalent
1 ft^3	6.24 UK gallons
1 ft^3	7·49 US/Canada gallons
1 ft^3/sec (cusec)	0·2832 m^3/sec
1 cusec	0·539 million gallons/day (mgd)
1 mgd	1·8 cusecs
6 ft (1 fathom)	1·828 m
1 international nautical mile	1·852 km
1 knott (international)	1·852 km/hour

Index

Books published by
Fishing News Books Ltd

Free catalogue available on request

Advances in fish science and technology
Aquaculture practices in Taiwan
Aquaculture training manual
Aquatic weed control
Atlantic salmon: its future
Better angling with simple science
British freshwater fishes
Business management in fisheries and aquaculture
Cage aquaculture
Calculations for fishing gear designs
Commercial fishing methods
Control of fish quality
The crayfish
Culture of bivalve molluscs
Design of small fishing vessels
Developments in fisheries research in Scotland
Echo sounding and sonar for fishing
The edible crab and its fishery in British waters
Eel culture
Engineering, economics and fisheries management
European inland water fish: a multilingual catalogue
FAO catalogue of fishing gear designs
FAO catalogue of small scale fishing gear
Fibre ropes for fishing gear
Fish and shellfish farming in coastal waters
Fish catching methods of the world
Fisheries oceanography and ecology
Fisheries of Australia
Fisheries sonar
Fishermen's handbook
Fishery development experiences
Fishing boats and their equipment
Fishing boats of the world 1
Fishing boats of the world 2
Fishing boats of the world 3
The fishing cadet's handbook
Fishing ports and markets
Fishing with light
Freezing and irradiation of fish

Freshwater fisheries management
Glossary of UK fishing gear terms
Handbook of trout and salmon diseases
A history of marine fish culture in Europe and North America
How to make and set nets
Introduction to fishery by-products
The lemon sole
A living from lobsters
Making and managing a trout lake
Managerial effectiveness in fisheries and aquaculture
Marine fisheries ecosystem
Marine pollution and sea life
Marketing in fisheries and aquaculture
Mending of fishing nets
Modern deep sea trawling gear
More Scottish fishing craft and their work
Multilingual dictionary of fish and fish products
Navigation primer for fishermen
Netting materials for fishing gear
Ocean forum
Pair trawling and pair seining
Pelagic and semi-pelagic trawling gear
Penaeid shrimps — their biology and management
Planning of aquaculture development
Refrigeration on fishing vessels
Salmon and trout farming in Norway
Salmon farming handbook
Scallop and queen fisheries in the British Isles
Scallops and the diver-fisherman
Seine fishing
Squid jigging from small boats
Stability and trim of fishing vessels
Study of the sea
Textbook of fish culture
Training fishermen at sea
Trends in fish utilization
Trout farming handbook
Trout farming manual
Tuna fishing with pole and line